SAVING THE ST. JOHNS RIVER

An Angler/Conservationist's Quest To Restore and Protect the Historic and Majestic St. Johns River

To:
Mark, Heidi + Julian.
Hope you enjoy reading all the efforts of my husband.
We spent 25 yrs in this effort
Love to you
Always
Leroy + Joyce

Written by:

Leroy Wright

One Person Made a Difference

First published by Dog Ear Publishing
4010 W. 86th Street, Ste H
Indianapolis, IN 46268
www.dogearpublishing.net

ISBN: 1-59858-239-9
Library of Congress Control Number: 2006936659

This book is printed on acid-free paper.

Printed in the United States of America

For Someone Very Special…

To my lovely wife Joyce, thank you Sweetheart for your enduring support. More than anyone, you understand my passion and commitment to restore and protect the St. Johns River for future generations. With my love, I dedicate this book to you.

CONTENTS

	Acknowledgements	*vii*
	Introduction	*ix*
Chapter I	*A Synopsis—Past History to Present*	*1*
Chapter II	*Beginning of the Journey and Beyond*	*19*
Chapter III	*Upper Basin Restoration Project*	*31*
Chapter IV	*Restoring Lakes Hell N' Blazes and Sawgrass*	*45*
Chapter V	*Proposed Sabal Hammocks Project*	*60*
Chapter VI	*Restoring the Ocklawaha River*	*81*
Chapter VII	*Private Lands vs. Sovereign (Public) Lands*	*100*
Chapter VIII	*Transforming Duda Ranch into New City of Viera*	*121*
Chapter IX	*The American Heritage River Initiative*	*137*
Chapter X	*Fruits of My Labor*	*155*
Appendices		
Appendix 1	*Upper Basin Boundary Map—St. Johns River*	*169*
Appendix 2	*Three Basins' Boundary Map—St. Johns River*	*170*
Appendix 3	*American Heritage River Designations—Nationwide*	*171*
Appendix 4	*River Lakes Conservation Map*	*172*

ACKNOWLEDGEMENTS

In 1985, Mr. David (Dave) Cox, a marine biologist with the Florida Freshwater Fish and Game Commission, played a key role in my decision to organize SAVE St. Johns River, Inc. SAVE is an acronym for Sportsmen Against Violating the Environment. Dave's knowledge of a troubled river stirred this writer into action. Thank you my friend, for your dedicated support over the past 21 years.

Mr. Maurice Sterling, Upper Basin Restoration Project Manager for the St. Johns River Water Management District, provided this writer an understanding of this massive restoration project. Thank you Maurice for your dedicated work in restoring the Upper Basin. I am privileged to have such a friend.

For over 21 years, Bill Sargent, Outdoor Writer for Florida Today, published many articles on our coalition's bass fishing tournaments/fundraiser events. Bill's timely articles were key to increasing public awareness, membership, and support for our coalition's mission to restore and protect a magnificent river. Thanks you Bill for your support over the years.

Former Governor Lawton Chiles and the Cabinet directed an investigation and determination of potential sovereignty lands on a proposed development on the St. Johns River in Brevard County. A special thanks to these state officials, and to our coalition's legal team: Attorney Tom Tomasello, of Tallahassee, Environmental Consultant Dennis Auth, of Jacksonville, and Water Quality Expert Dr. Forrest Dierberg, formerly of Melbourne.

The Brevard County Commission supported our coalition on numerous issues regarding restoration and protection of the St. Johns River. For their consistent support, I especially thank Commissioners Nancy Higgs, Sue Carlson and Truman Scarborough.

Former Jacksonville Mayor John Delaney filed the application to designate the St. Johns River as an American Heritage River. Pat Poole, former Melbourne city council member, obtained resolutions from 14 municipalities in Brevard County which was significant in the process of obtaining this special federal designation. A special thanks to these dedicated supporters for their work to restore the St. Johns River.

I extend my personal appreciation to the following individuals, who in their own special way, provided assistance to SAVE St. Johns River, Inc. over the years. I apologize if I missed anyone:

Lothian Ager	Glenda Busick	"Scooter Creel"	Fred Cross
Diana Sawaya-Crane	Bill Daniel	Jack Eckdahl	Margarita Engle
David Godfrey	Larry Gleason	Pricilla Griffith	David Guest
Hector Herrera	John Hankinson	Dan Hite	Carol Hupfer
Bob Eisenhaur	Tom Lawton	Charles Lee	Jack Maney
Jack Masson	Robert Montgomery	Dr. Al Mills	Carol Pope
Roger Patton	Steven Robinson	Bill Rucker	Wanda Rucker
John Richardson	Neil Ridenour	Mike Stevens	Jack Stanley
Doug Sphar	Mary Sphar	Ron Taylor	Toby Tobin
Rick Thrift	Larry Webber	Joyce Wright	

The following books present a variety of historic, environmental, and sometimes comical events. I recommend them for your reading enjoyment:

1) Travels of William Bartram, Dover Publications, Inc., New York, 1955;
2) The Riverkeepers, Joe Cronin & Robert F. Kennedy, Jr., Scribner, New York, 1997;
3) Voice of the River, Marjory Stoneman Douglas, Pineapple Press, Inc., Sarasota, Fl 1987;
4) River of Lakes, A Journey on Florida's St. Johns River, Bill Belleville, Georgia Press, 2000;
5) Tales from a Florida Fish Camp, Jack Montrose, Pineapple Press, Sarasota, Fl., 2003;
6) The St. Johns—From Marshlands to the Atlantic, Harrell & Geiger, Fla. Historic Society.

INTRODUCTION

The St. Johns River is one of several in the world that flows in a south-to-north direction. From the headwaters in Indian River County, this slow-moving river drops only 33 feet in its 310-mile journey to the Atlantic Ocean, at Mayport, north of Jacksonville. The normal, but slow moving current is responsible for the river being referred to as "a lazy river".

In the more recent history of the St. Johns River, the Timucua and Seminole Indians were among the earlier settlers. Subsequently, Spanish, French, and European explorers would claim ownership throughout the reaches of the river and its chain of lakes.

Upon their arrival in the 1500s, the Spanish routed the Timucua from the river. The Spanish explorers named the river: "Rio de Corriente", meaning "River of Currents". Following the decimation of the Timucua Indians, the Seminole Indians migrated to the St. Johns River, relocating from Georgia and Alabama. The Seminoles named the river: "Welaka", meaning: "River of Lakes".

Later, the Spanish defeated a short lived French occupation and renamed the river: "San Mateo" in honor of a feast day for a catholic saint. In the early 1600s, however, the river's name was changed yet again to: "Rio de San Juan". In the 1760s, when the British took possession of the river, its name was translated into English, meaning: "St. Johns River". Finally, in 1821, the United States purchased the river from the Spanish government, retaining its name as we know it today—St. Johns River.

My dad introduced me to fishing when I was 6 years old. I thanked him many times for teaching me how to fish. I lovingly remember him each time I have the opportunity to "wet a line". As my own kids began school, I taught each of them the fun in fishing. One might say: "it's a family tradition".

In late December 1958, I visited my sister in Rockledge, Florida. I planned a fishing trip on the St. Johns River the following week. I departed the boat ramp as the sun began to appear on the horizon. One hour later, I hooked and landed a seven-pound largemouth bass. I had hooked the bass, but it was I who really got hooked. I decided I wanted to live near this magnificent river.

The next day I applied for employment with the Martin Company (now Lockheed-Martin Corp) at Cape Canaveral, Florida. I was offered a position and asked if I could report for work the following day. I informed the Personnel Manager that I was in Florida on vacation and would not complete my active duty in the U.S. Air Force until 17 January 1959. The company held the position for me; my reporting date was set for 19 January 1959. I enjoyed an exciting 30-year career in the Aerospace Industry. I retired as Chief, Design Engineering & Support in January 1989 at age 56. I looked forward to great fishing and boating on the St. Johns River and its many lakes.

For 48 years, the Upper Basin (the first 111 miles) of the St. Johns River has been my back yard. During these many years, I explored and fished every lake throughout this magnificent river, from the first lake (Hell N' Blazes) to Green Cove Springs, near Jacksonville. In the late 1970s and early 1980s, I experienced a decrease in my fishing success while observing a definite increase in poor water quality throughout the Upper Basin, especially following major storm events. I became very concerned about the health of the St. Johns River.

In late 1983, I began plans to create an organization that would address the problems of a dying river. I needed to form relationships with anglers and hunters who enjoyed spending time on the river. I approached the matter by starting a new business, which I named: "Hunt N Fish". I located my business venture on SR 520, a high-traffic highway in central Brevard County, only 4 miles from the St. Johns River.

On week-ends, I conducted the largest bass-fishing tournaments ever held in this area of the St. Johns. In numerous tournaments, over 100 boats (200 anglers) would compete for cash and prizes. I was able to witness the "catch rate" for largemouth bass. Over the next two years, the data clearly supported my conclusion that the bass fishery was diminishing. Years would pass before the Florida Freshwater Game and Fish Commission reduced the daily bag limit for bass from 10 per day to 5. Since I was still employed at Cape Canaveral (through 1988), my son, Jimmy, who was very knowledgeable in all aspects of fishing and hunting was my first store manager. The successful bass-fishing tournaments had placed this writer in a position to meet hundreds of anglers and hunters. These individuals would play key roles, both physically and financially in the new coalition I was now prepared to form.

In October 1985, at a public meeting I had arranged via the news media, I addressed my concerns about the river's health. Seventy-five attendees were in agreement with me. A coalition comprised of one homeowner association and nine fishing/hunting clubs was formed. This writer was elected President. I suggested the name "Sportsmen Against Violating the Environment (SAVE) for our newly formed coalition. At the next meeting, I recommended the acronym SAVE be supplemented with "St. Johns River" which would identify our specific mission. The new name, "SAVE St. Johns River" was adopted.

The book describes our organization's work with local and state government officials to bring about enforcement of existing rules and changes in other rules, when necessary, to restore the river's health. By the early 1990s, our coalition was incorporated. SAVE St. Johns River, Inc. now included 30 organizations, with a total support base of over 3,000 residents in Brevard, Indian River, and Orange counties.

Within the chapters of this book, this writer will reveal how an historic river, teaming with an abundant fishery and many types of wildlife, almost became a death trap for these species. For many years, a small number of ranchers and farmers with title to thousands of acres of the Upper Basin marsh and headwaters, were more committed to profit than protecting the river's fishery and wildlife. The reader will discover how this writer worked with our support base and government agencies in addressing many of the water quality issues on the river.

This writer's engagement with local and state officials resulted in the state purchase of 14,137 acres of the 38,000-acre Duda Ranch in Brevard County. The purchase included 14 miles of the east shoreline of the ranch. Much of the 14,137 acres was originally planned for future development as part of the new city of Viera.

Another significant win for the river resulted when this writer joined with local, state and federal officials in securing the federal designation of the St. Johns River as an "American Heritage River" by former President Clinton. My wife (Joyce) and I received a personal invitation from the White House, to join the President in North Carolina where he would announce the 14 rivers selected to receive this special designation. There were 126 rivers nominated nation wide.

The St. Johns River boundary is divided into three basins: the Upper Basin, Middle Basin, and Lower Basin. Until 1992, my primary focus was dedicated to the Upper Basin. However, our coalition became an affiliate member group of Florida Defenders of the Environment, located in Gainesville. We support removal of Rodman Dam, located on the Ocklawaha River, a major tributary of the St. Johns River. A free flowing Ocklawaha would once again resumes its natural flow, returning this canopied paradise to a river noted for its beauty and great outdoor recreation.

In the Upper Basin, for over 50 years, agricultural pump discharges from ranches and farms significantly contributed to major fish kills. In the Middle Basin, similar conditions exist, in addition to poorly planned development in some areas of the basin. In the Lower Basin, septic tank malfunctions and unsustainable development render poor water quality in many areas of the basin. Lesions appear on many fish species due to heavy metal deposits. Common among all three basins are the recurring number of fish kills, resulting from low levels of dissolved oxygen (poor water quality).

On a more positive note, the Upper Basin is experiencing improvements in water quality as the massive Upper Basin Restoration Project nears completion. Farm runoff from agricultural operations is now treated in large holding areas prior to release into the river. In addition, by issue of Consent Orders, agricultural operations now retain the first inch of rain on-site via retention reservoirs. As the waters of the Upper Basin flow north, the Middle and Lower Basins should experience improvements in water quality.

On the St. Johns River, nature reveals its healing power in many ways. On my visits to the river, I still experience such sightings as: a bald eagle perched atop a cypress tree; various species of wading birds feeding in the shallows; river otters playing along the shoreline; a blue heron "fishing" for a meal; an alligator "bellowing" as mating season begins; wild plants blooming along the river banks and in the marsh areas; deer crossing the vast shallows of the marsh; the splendor of a sunrise over the calm waters, or an orange-gold sunset as darkness approaches. On numerous occasions, I stop fishing and sit quietly observing the bounty of nature's gift to all of us.

I revitalize my own spirit with frequent trips to the river. On some of these trips, I provide elected officials the opportunity to view for themselves the need to restore this great river. I have provided these "educational tours" for the past 21 years.

In writing this book, I provide the reader a look at this writer's lifestyle. At the conclusion of each chapter, except the final one, I think you will enjoy a few of my memorable fishing stories. Did I say fishing stories conclude each chapter? Fishing and observing nature have been my way of relaxing since I was a child. After completing each chapter, I would put away the writing pen and head to the St. Johns River.

This writer is humbled by the special recognition I have received for my volunteer work on the St. Johns River. More important, however, the essence of my work will be fulfilled when future generations have the opportunity to experience the splendor of God's gift to all who take their own personal journey on this great river. I trust the book will inspire the reader and other "volunteers" to carry the torch to the next generation. The St. Johns River belongs to each of us. Be a steward of the river. As the reader will discover, **one person made a difference.**

CHAPTER I—

PAST HISTORY TO PRESENT

This chapter presents a brief review of how man's attempt to change an historic river's natural flow almost destroyed the headwaters of the St. Johns River. The headwaters demise seriously affected the first five lakes on the St. Johns River. These lakes include Lakes Hell N' Blazes, Sawgrass, Washington, Winder and Poinsett. The distance from Lake Hell N' Blazes to Lake Poinsett is approximately 30 river miles. These lakes are a significant part of the Upper Basin (refer to Appendix 1, Map of Upper Basin Boundary). My years of enjoyable experiences "in my back yard" began to fade by the late 1970s. I will share with the reader how I established a grassroots coalition determined to stop the river's slow death. Chapter II outlines the formative years of our coalition entitled "Beginning the Journey and Beyond". The subsequent chapters reveal specific successes and some challenges this writer is committed to continue for one reason only; to restore the health of the St. Johns River.

William Bartram recorded his account of travel to the St. Johns River in his "collector book", Travels of William Bartram, first published in Philadelphia in 1791. He was a naturalist, a botanist, and explorer. Bartram first traveled along the St. Johns in the early 1760s. His curiosity and love of nature enabled him to document his experiences on the river, and deal with the early settlers, especially the Indians. He also observed wildlife and plants along the river. Bartram's father, John of Philadelphia was also a botanist who had traveled east Florida, collecting many specimens. John Bartram established the first botanical garden in America. In a city park in Philadelphia, part of the garden has been preserved.

The St. Johns River served as a highway for transportation in its early history. Before the railroad tracts were built, steamboats were popular, ferrying merchandise and travelers by the mid to late 1800s. The St. Johns had an abundance of wildlife, including fish, alligators, turtles, frogs, ducks, wading birds, river otters, deer and other species. Many of the early settlers derived a living from the vast assets of the river. Orange groves were present at the time of Bartram's travels. Many commercial boats made routine trips as far south as the Cocoa/Rockledge area. To accommodate the loading and unloading of people and supplies from the boats, in some instances horse-drawn wagons would be placed into the shallow waters of the river.

In 1910, three men were fishing in the Puzzle Lake section of the St. Johns River, south of present day SR 50. They had fished for several days, during which time they caught over 2,200 pounds of large-mouth bass (I have seen the photo of their catch). Many of the bass weighed over 10 pounds each. No law regulated how many bass per day one could keep. However, no refrigeration existed at the time. Perhaps the community enjoyed a three-day picnic, sponsored by these hearty anglers.

Also in 1910, New Zealand immigrant E. Nelson Fells bought 116,000 acres of the "swamplands" of the Upper Basin of the St. Johns River. The town of Fellsmere was born. Mr. Fells paid $63,000 dollars for the property. He would undertake a massive engineering project to drain the "swamp". He named the project Fellsmere. The original plan for the property was to establish a project that would be bigger than the city of Miami. Fellsmere Inn opened in 1911. The Inn was a centerpiece of the new town.

In 1916, a major flood dropped 16 inches of rain within 48 hours into the area. Homes were flooded; many settlers left the area by boat, never to return. That same year, the town of Fellsmere was sold for $1.00. In the 1930s, a Sugar Mill opened. The mill operated until the 1960s. In 2004, the Inn incurred significant damage from multiple hurricanes. The town leaders decided the Inn would be leveled. Before it was to be destroyed, Mr. Fred Vanderveer, a real-estate investor from the Florida Keys purchased the Inn from Margaret Freeman for $300,000 dollars; he plans to restore the historic Inn. Restoration includes a restaurant and hotel for visitors, especially those who fish the nearby and nationally known waters of the Stick Marsh, a 6,000-acre state-owned reservoir noted for its big largemouth bass population.

In the early 1900s, cattle and citrus were becoming a significant investment on the part of the early settlers in the Upper Basin region. The ranchers and farmers were the people who worked the land and put food on the table for the early residents. Most of these landowners continue to grow crops and raise beef cattle on their property. Agricultural operations significantly contributed to the economy along the headwaters of the St. Johns River.

The St. Johns River and its floodplain provided these early settlers very fertile soils with an ample water supply. These pioneers took full advantage of the river and its lush marshland. Most all cattle operations were either located or relocated to the St. Johns River floodplain. One such story was reported by Chuck Reed, native born and great-grandson of a central Brevard pioneering family. Chuck reported that many of the local pioneers on Merritt Island made their living from citrus and cattle operations. Cattle were allowed to graze freely in uncleared areas of the island. Such areas were known as open-range lands.

The story continues—in 1917, the first bridge over the Indian River to Cocoa was built. Prosperous citrus growers and a waive of settlers on Merritt Island soon came into conflict with the hungry cattle, which enjoyed feeding on young citrus trees and vegetable gardens of the new homesteaders. Cattle owners decided to move their herds to the vast St. Johns River basin, west of Cocoa. It was possible; after all, there was a new bridge to the mainland of Cocoa.

Mr. Platt, a pioneer rancher in West Melbourne agreed to organize a cattle drive to relocate the herd. Brevard County government decided a $20,000 dollar bond had to be posted to ensure the bridge from damage. According to the story, Mr. Platt said he did not have that kind of cash or time to find it, but if the County would make a check for that amount for his bond, he would sign it and get the cattle drive on the way. The County agreed; the cattle passed over the bridge and through Cocoa without incident. Mr. Platt, in turn, stated to County officials: *"Well, that's that, so now you can just tear up that check"*.

In 1920, the big town of Fellsmere had over 800 residents. There were 749 automobiles registered in Brevard County. In 1922, a road was built across the St. Johns River flats from Mims, in Brevard County, to Oviedo, in Seminole County. J. J. Parrish proudly announced that he had driven his Buick from Mims to Orlando in only 2 hours and 30 minutes.

In 1925, a seven-mile dike was constructed across the St. Johns River marsh from the Fellsmere area. The dike would provide a road to "Kissimmee County". Fellsmere was predicted to be a "boom-town". A wooden bridge was constructed as part of the new dike "road" to allow marsh waters to flow under the bridge, continuing its journey into the St. Johns River. Legend has it that a local man started a relationship with a woman from the Kenansville area. He utilized the bridge to visit her. Subsequently the bridge mysteriously burned to the ground, thus ending the romance.

The historic Fellsmere Estate Building was constructed in 1926. It was used as the land sales office for the Fellsmere Estates Corporation. Following earlier flooding of the area, coupled with the depression, which started in 1929, the Florida land boom crashed, along with the stock market. The Fellsmere Estate Corporation closed its doors. In the 1930s, the building became the headquarters for the Florida Crystal Sugar Company, as sugar was the town's main commodity through the early 1960s. That industry would soon disappear as sugar plantations expanded in south Florida.

Subsequently, the building became a municipal building with the City Council meetings held in the large interior room that is now the main dining room of the present day "Marsh Landing" restaurant. In 1995, former Indian River County Commissioner Fran Adams bought the property at a public auction for historic restoration. Indeed, she has restored the building to near original condition. The building has elegance unmatched by modern-day buildings. Joyce and I recently enjoyed a great lunch on the premises. I highly recommend a trip to the "old town" restaurant soon. One can enjoy real home style meals in an atmosphere rarely found in other public eateries. A visit will place you in an earlier Florida boom and bust time.

Marsh Landing Restaurant in Fellsmere (Indian River County)

Landowners along the St. Johns River held title to vast tracts of land; some owned thousands of acres of what is today known as the headwaters of the St. Johns River. Two examples: Deseret Ranch stretches 305,000 acres through parts of four counties. The Mormon Church owns these lands. Their land includes thousands of acres bordering the west side of the river, while the Duda Ranch property, originally at 38,000 acres borders the east side of the river in Brevard County. As stated earlier, a small number of individuals owned most of the land surrounding the Upper Basin of the St. Johns River.

Dredging and Diking the Upper Basin Marshlands

As World War II was nearing an end, these lands were being evaluated for their potential to "feed the people". Massive drainage projects were underway. The federal government provided support to continue draining the marsh on the east and west sides of the river. Actual draining began slowly in the early 1900s. By the 1950s and 1960s, dredging and diking became commonplace in the Upper Basin. The federal Clean Water Act was not enacted until 1972; therefore, much of the dredging and ditching occurred with little or no opposition.

Environmental protection rules were either non-existent or ignored for many years. It is my opinion that the mind-set of the landowners was simple: a massive river was available that could easily handle pump discharges from the farms. Nothing was hidden from public view; apparently, the agricultural folks were not breaking any laws. It was a way of life on the farm, or ranch. Since they held title to the land, some grew crops well out into the marsh. The soil was fertile. Dikes kept the high water outside. This writer probably enjoyed some of the vegetables produced on these farms.

The St. Johns River Water Management District, hereinafter referred to as the SJRWMD, was created in 1972. Prior to that time, the Central Florida Flood Control District was responsible for monitoring the river's water problems. Those problems primarily dealt with flood control—protecting the upland landowner's property. The Florida Freshwater Game and Fish Commission was primarily responsible for enforcement laws relative to bag limits, fishing license, and poaching.

In 1984, Deseret Ranch filed a lawsuit against the SJRWMD. Deseret claimed the ranch was exempt from rules on a dredging project. Deseret was performing the dredging to keep their upland property from flooding. The SJRWMD stated a state permit was required; all work was stopped. Deseret was dredging canals and using the excavated soil to build up dikes along the river in south Brevard County. The case was decided in favor of the SJRWMD at the Fifth District Court of Appeals. The decision was precedent setting in environmental law and increased the SJRWMDs authority in regulating farms in the river's headwaters.

In the spring of 1986, Ed Vosatka was a biologist with the Florida Freshwater Game and Fish Commission. On one of his many trips on the river, Ed looked out at the slow moving water and shoreline, and commented: *"As beautiful as the river and the surrounding land is, the area is well on its way to destruction...the willows, maples and other trees are signs the marsh is drying up"*. At the time he spoke those words, he was right. Dave Cox, also a biologist for the same agency echoed similar statements. Dave stated: *"The River is dying"*. These knowledgeable officials spent much of their time on the water. **Except for an aggressive Upper Basin Restoration Project, I would be describing a paradise lost.** Read the story of the world's largest (at the time) restoration project: Chapter III, Upper Basin Restoration Project.

By the 1980s, agricultural interests had drained and diked over 100,000 acres of marsh and wetlands. Bulldozers and dragline operators created new dikes throughout the Upper Basin marsh areas. Large capacity pump stations located at selected diked areas offloaded massive volumes of poor quality water into multiple canal systems leading directly into the St. Johns River. The polluted water discharges were now a part of the river, moving slowly through Lakes Hell N' Blazes and Sawgrass, enroute to Lake Washington and all points north.

SAVE Initiates Action to Protect a Fragile River

In 1988, our coalition, SAVE St. Johns River, hereinafter referred to as SAVE, requested the SJRWMD and Florida Freshwater Game and Fish Commission (GFC) undertake a feasibility study to restore Lakes Hell N' Blazes and Sawgrass, the first two lakes in the headwaters of the St. Johns River. Read about missed opportunities, but hope eternal, to restore these lakes: Chapter IV, Restoring Lakes Hell N' Blazes & Sawgrass.

In June 1990, SAVE requested approval from the Brevard County Commission and the SJRWMD to adopt the meandering waters in the Upper Basin of the St. Johns River. Approval was granted. Seventeen "Adopt a River" signs were provided by the SJRWMD. This writer had previously obtained authorization from private fish camp owners and marinas to place the signs at boat ramps along the adopted stretch of the river. Signs were also placed at all Brevard County park boat ramps on the St. Johns River.

Also in June 1990, SAVE filed a legal challenge to stop a proposed development called Sabal Hammocks. The project was to be located behind an 18,000-foot illegal dike on the northeast shoreline of Lake Poinsett in Brevard County. Lake Poinsett is the fifth lake north of the headwaters of the St. Johns River. This would be the start of a 15-year plus journey through the courts. The Florida Governor and cabinet would address the issue on three separate occasions. Read this most interesting story: Chapter V, Proposed Sabal Hammocks Project.

News Media—A Voice on the River's Demise

In the summer of 1992, an Orlando Sentinel article addressed another massive fish kill on Lake Winder in Brevard County. Over 235,000 fish had suffocated. Joyce and I visited this "graveyard" to witness and videotape the scene. Upon returning home, I turned on the local NBC TV affiliate station for the noonday news. As I watched in amazement, the news anchor produced a video tape of an early morning interview with a Duda Ranch representative. The Duda official stated there were no fish killed in their canals leading into Lake Winder. The interview was conducted on the very same morning Joyce and I videotaped thousands of dead fish in the Duda canals and lake.

As soon as the news segment concluded, I took our videotape, which displayed the time and date, directly to the local NBC TV affiliate station in Cocoa Beach. I waited while our video was reviewed by the news staff. That evening, on the 6 PM news, our video was aired. Vultures were shown feeding on the dead fish along the banks of the Duda canal, which runs directly into Lake Winder. Other scenes displayed floating dead fish in the canal and well out into Lake Winder. The record was set straight. The Duda Ranch was just one among other ranches doing the same thing; draining their lands to protect their interests (cattle, citrus, sod and row crops).

In October 1992, the Orlando Sentinel published a telling story of the problems with the entire St. Johns River entitled "Killing the St. Johns". (See map provided in Appendix 2, depicting the boundary of each basin of the river: Upper, Middle and Lower.)

In the Upper Basin, the Sentinel reported massive fish kills and poor water quality were the culprits; in the Middle Basin, more fish kills, algae blooms, and declining water quality; in the Lower Basin, fish with tumors and ulcers, high levels of sewage, heavy metals, and pesticides were present. Common among all three basins was poor water quality, which resulted in diseased or dead fish throughout the 310-mile length of the St. Johns. One SJRWMD official commented, *"What we are seeing in the river is a clear red flag, the canary in the mine shaft. We have a significant problem"*.

The Sentinel article concluded by addressing Lake Jessup in the Middle Basin. The lake was formerly famous for its good fishing (this writer enjoyed the lake's fishing in the 1970s). It is now a 10,000-acre mud-hole with sediment at least two feet deep, covering the lake bottom in most places. Sewage and farm runoff poured into the lake until the early 1980s. The lake's nutrient levels were found to be higher than 90 percent of all other lakes in Florida. I understand funding has finally been allocated to begin restoring the lake. However, Lake Jesup has a unique characteristic; there is no "outflow" from the lake as occurs on all other lakes on the St. Johns River. The natural configuration of the lake supports only a single opening, therefore little current exists. Only winds can assist in any recharge of the waters.

The December 1992 issue of the national publication "Bassmasters" with over 650,000 members (38,000 in Florida) published a significant three-page story on the negative effects of agricultural pumping operations in the Upper Basin, especially as to fish kills. As a member for many years, I looked forward to each issue, especially the environmental articles written by Senior Writer Robert Montgomery. I contacted Mr. Al Mills, Director of Natural Resources. This writer made a personal request for Mr. Montgomery to travel to Florida for a major story on the environmental devastation of the St. Johns. After briefing Mr. Mills on the continuing loss of a great river's fishery, I was put in contact with Mr. Montgomery. We arranged a time and place to meet at the river the very next week.

Using my boat, we proceeded to a large pump station at a Duda Ranch dike near Lake Winder. Timing was excellent; the pump was in operation. A chocolate colored water column was exiting the pump station into a Duda Ranch canal that offloaded into the lake. A small distance from the pump outflow, bubbles were rising to the surface, producing a stench that smelled of rotting eggs. The odor was caused by methane gas rising to the surface; a result of the stirring up of sediment from the bottom of the canal. We visited several other areas of the dikes and the lake. Mr. Montgomery was constantly asking questions and writing down his observations. I reminded him that what he was witnessing occurred every time a significant storm event passed through the river basin. Indeed, the article depicted how the public was losing a great fishery.

The most significant story ever published on the St. Johns River by Florida Today occurred on 17 December 1992. A 14-page special edition entitled "Blueprint Brevard—The St. Johns River" was indeed informative and provided every interest group an opportunity to express their viewpoint about the river and its future. Scientists, biologists, ranchers, conservationists, anglers, nature lovers, government officials, etc. all documented their views. Page 11 was dedicated to tell SAVEs story.

Appropriately entitled: *"To Save the St. Johns River"*. I was provided an opportunity to inform thousands of citizens throughout the Upper Basin of the river's problematic future. In my interview for the special edition, a sub-heading read: *"A Talk with SAVEs Founder"*. My views regarding the river's future culminated in a simple statement—*"It is time all parties come to the table and work out a plan to restore the St. Johns River"*. Following publication of this special edition, SAVEs prominence significantly increased. Our membership exploded over the next year to 30 organizations. Our support base increased into the thousands. The Florida Wildlife Federation, local Homeowner Associations, Condo Associations, and numerous outdoor groups joined with SAVE to demand the river's health be restored.

SAVE Expands Support to Other Groups

In late December 1992, I was contacted by a group of concerned citizens who utilized the Tosohatchee State Reserve. The Reserve borders the St. Johns River for 19 miles, north of SR 520 in Orange County. The state parks agency had posted notice that the state was considering closing the Reserve. Joyce and I joined the group at the gate entrance early the next morning. After a tour of the vast 28,000-acre property, SAVE joined their effort to keep the Reserve open to the public. Following an exchange of letters and phone calls with the state park's director in Tallahassee, an agreement was reached to leave the Reserve open to the public. The park is an excellent area for horseback riding, hiking and just enjoying one of Florida's true wilderness areas.

In 1993, SAVE became an alliance member of the Florida Defenders of the Environment (FDE). FDE is located in Gainesville, Florida. We continue to support their effort to remove Rodman Dam and restore the Ocklawaha River, a tributary of the St. Johns River. The dam was part of the old Cross-Florida Barge Canal project that President Nixon cancelled in 1971. Read about this issue: Chapter VI, Restoring the Ocklawaha River.

Florida Wildlife Federation Service

In September 1993, I was honored by the Florida Wildlife Federation (FWF) as "Conservationist of the Year" at their annual Award Banquet held at the Greenleaf Resort in Haines City, Florida. Joyce, the children, their spouses, and special friends were invited. All of us celebrated the event. I am very proud of the "Eagle" statue presented to me that evening. Joyce was instrumental in this award. She provided insight and documentation during the nomination process, leading to this prestigious award.

In 1994, I was nominated to serve on the Board of Directors for the Florida Wildlife Federation. Mr. Larry Gleason, Vice-President of SAVE, nominated me for the position. I was elected as Regional Director for Central Florida for four years. In September 1998, I resigned my position to work with government officials, conservation groups, the media, and citizens in securing the St. Johns River's designation as an American Heritage River.

State Takes Over SAVE Lawsuit

In December 1995, Governor Lawton Chiles and the Cabinet voted 7-0 to proceed with a quiet title claim against David A. Smith, owner and developer of the proposed Sabal Hammocks Project. The state would claim ownership of all sovereign lands below the Ordinary High Water Line (OHWL) of Lake Poinsett. Read how SAVE approached this issue: Chapter VII, Private Lands vs. Sovereign (Public) Lands.

SAVE Participates in Riverwide Clean-up

The Cocoa Tribune, a part of the Florida Today newspaper, reported on 18 February 1996 about an upcoming event regarding the clean-up of the entire St. Johns River. The event was scheduled to occur on 2 March 1996. The SJRWMD organized the event and held the initial meeting at its headquarters in Palatka. The event was well coordinated throughout the 14 counties in the watershed. SAVE participated in the river-wide clean-up. This writer volunteered to serve as the Brevard County Chairman for the Upper Basin. Many businesses donated bags, gloves, water, first aid kits, etc.

Thousands of supporters from throughout the river's length pitched in to perform the clean-up. With the assistance of hundreds of volunteers, plus county dump trucks, front-end loaders, and Western Waste's agreement to pick up the trash, over 300,000 pounds of litter was removed from the Upper Basin alone. Included was over 100,000 pounds of litter removed from an illegal "dump" in the marsh of the river, immediately north of US Highway 192 in south Brevard County. The event was an overwhelming success and has become an annual affair However, due to the logistics involved, each basin now schedules the date and time for their event.

Weather Shelters Replace Illegal Camps

Catfish Hotel; an old camp St. Johns River (removed by order of SJRWMD)

In 1996, I represented SAVE in meetings with the SJRWMD regarding the removal of illegal camps along the Upper Basin of the river. This writer recommended the SJRWMD construct weather shelters which would offer refuge from storm events. Illegal camps had been built along this part of the river over the past 50 to 75 years. Most were built on state lands by hunters and anglers to provide a camp for family and friends, as well as a refuge during inclement weather. The SJRWMD issued notice that the camps were to be removed as they were environmentally damaging to the river and could be a liability issue if someone was injured at one of the camps. An agreement was reached to construct a sufficient number of weather shelters as the illegal camps were removed. Over the years, a number of the old camps were destroyed by fire. The weather shelters are not for use as a campsite. Campsite areas (pads) were constructed in a number of areas to provide camping opportunities along the river and its lakes.

New Weather Shelter, Middle River (typical of many constructed in the Upper Basin)

SAVE Opposes New Power Plant

In late 1996, the Oleander Power Company wanted to build an 800 mega-watts power plant near the SR 520 and I-95 intersection. The plant would be approximately 5 miles east of the St. Johns River. Some experts claimed the pollutant load from the plant's stack-pipes would drift westward toward the St. Johns River since our winds are predominately from the southeast. SAVE was approached by one of our member groups, the Lake Poinsett Homeowners Association. After much discussion, I suggested the homeowners form a group of citizens to address the county commission. I further stated that I would represent SAVE to the commission, but to have another separate group present would be a great asset.

The group was formed. Known as Brevard Citizens Against Pollution (BCAP), they also had very vocal support from one of SAVEs condominium association groups. BCAP members were outspoken at subsequent county commission meetings. A six-month moratorium was put in effect by the commission to provide time for the board to decide how to proceed with the applicant. In the end, the county attorney reminded the commission that the county had no ordinance that addressed "heavy" construction; therefore, the county would not be in a favorable position to deny the application. SAVE provided financial assistance to BCAP in retaining an attorney. Indeed, SAVE had the support of a majority of the county commission members, but the commission did not have legal standing to win in a court setting. After a one-year delay, the county approved construction of the plant. The plant provides power outside Brevard County on an as-needed basis.

Opposition—Exchange of Bull Creek Lands

More activity in 1996: I informed the Director of Planning and Land Acquisition for the SJRWMD of SAVEs opposition to trading 10,000 acres of the popular Bull Creek Wildlife Management Area for 8,800 acres of marsh located in the Upper Basin, west of the St. Johns River. The situation was similar to the Oleander Power Plant, in that I was contacted by Mr. Don Aycock, a member of a large hunting group from Brevard and Osceola counties. Mr. Aycock sought my advice on how the group should proceed to stop the exchange of 10,000 acres of prime hunting grounds at Bull Creek for 8,800 acres of marshlands. The Bull Creek Wildlife Management Area was also popular for wildlife viewing, hiking, etc. The loss of 10,000 acres of prime lands would leave only 13,471 acres for public use, much of which functions similar to a marsh system during the long summer wet season.

I encouraged Mr. Aycock to request the 100 plus members of the hunting group join in membership with SAVE. Together we would do our best to ensure the exchange did not occur. Friends of Bull Creek was established and became a SAVE member group. I worked with the management staff of the SJRWMD while our new alliance members attended all meetings involving the transaction. Finally, after much public opposition, and the fact that the lands in consideration were not equal in value, the SJRWMD declined an exchange of the properties.

President Clinton to Honor Special River Designations

In 1997, President Clinton signed an Executive Order whereby 10 rivers in the United States would be honored by designating each river as an "American Heritage River". Because of the close competition, 14 rivers received this prestigious designation. There were 126 rivers nominated for consideration by the White House Council on Environmental Quality, headed by Mr. Ray Clark. As stated earlier, I resigned my position as Central Florida Regional Director for the Florida Wildlife Federation to devote myself full-time in securing the federal designation of the St. Johns River as an American Heritage River. Chapter IX—The American Heritage River

Initiative describes how Joyce and I handled the excitement surrounding the selection of the St. Johns River as one of 14 rivers to receive this special designation. (Appendix 3 provides a list of the 14 rivers selected.)

SAVE Honored by Florida Wildlife Federation

In 1998, Mr. David Cox, a marine biologist with the Florida Freshwater Game and Fish Commission, nominated the SAVE organization to receive the Florida Wildlife Federation's "Conservation Organization of the Year" for our work in restoring the St. Johns River. SAVE won this award; a replica of a Bison is proudly displayed in our home. At this same event, Mr. Don Aycock, President of SAVEs affiliate group, Friends of Bull Creek, received the "Land Conservationist of the Year" award for their group's effort in resisting the exchange of 10,000 acres of Bull Creek. One other award was presented to Bill Sargent, Outdoor Writer for the Florida Today newspaper. Bill received the federation's Burk "Biff" Lampton Conservation Communicator of the Year award.

SAVEs Effort Culminates In Purchase of Significant Acreage

In 1999, at a cost of $24.8 million dollars, the state purchased 14,137 acres of low-lying lands of the Duda Ranch, bordering 14 miles of the east shoreline and floodplain of the St. Johns River in Brevard County. For years, the owners declined to sell any part of their 38,000-acre ranch. In this writer's opinion, the Viera Company was significantly encouraged to rethink their planning for a total build-out of the ranch property. Read the interesting account of how SAVE won the support of the Brevard County Commission in the Viera development order: Chapter VIII, Transforming Duda Ranch—City of Viera.

Deer Run Development—in St. Johns River 25-Year Floodplain

On 6 August 2001 (my 69th birthday), Florida Today Newspaper featured this writer in its Guest Column. The subject of my article—"Limit Development in the Floodplain". I addressed the difficult times experienced by residents of a recently built Deer Run subdivision in south Brevard County. I had watched TV coverage a few days earlier of a sad situation unfolding.

Unsuspecting homeowners had purchased homes in the 25-year floodplain of the St. Johns River. The homes and property were beautiful at the time of their purchase. However, a major storm event in the summer of 2001 caused nearby canals leading to the St. Johns River to overflow into the subdivision. The residents of Deer Run called upon county government and the SJRWMD for help. In a coordinated effort between the county and the SJRWMD, pumps were installed on the canal. Weeks passed before the water receded. The homes had been constructed at the required 100-year flood elevation. However, roads, driveways and outbuildings were flooded; bathrooms were unusable due to septic tanks back flowing into the homes. SAVE strongly disagrees with any development in the annual to 25-year floodplain of the St. Johns River. Brevard County's Comprehensive Plan currently permits only limited development in the annual to 25-year floodplain. I hope that the Deer Run saga will never play out again on the St. Johns River floodplain.

St. Johns River Alliance Organized

In 2003, the St. Johns River Alliance (SJRA) emerged out of the river's designation as an American Heritage River. The SJRA was approved as a 501(c) (3) non-profit corporation. The original steering committee became the Board of Directors for the new alliance.

As an original member of the steering committee, this writer had previously contacted Mayor John Delaney of Jacksonville, Chairman of the steering committee. I suggested he contact Brevard County Commissioner Sue Carlson and Indian River County Commissioner Ruth Stanbridge and request each of them serve on the steering committee. Both commissioners agreed to represent their respective counties.

The SJRA board members represent 14 counties along the St. Johns River. Over $4.6 billion dollars in needed restoration projects have been identified and prioritized by the SJRA throughout the 310-mile length of the river. Read the story of how the St. Johns River was awarded the designation as an American Heritage River; also how and why this initiative was transformed into the SJRA: Chapter IX, The American Heritage River Initiative.

Leroy Wright Recreation Area

In October 2004, I was contacted by the SJRWMD and requested to accompany staff officials on a visit to a site on the river, north of SR 520 that I had identified as a project for the SJRA. Joyce travels with me on most of my ventures on the river; she accompanied me on this "scheduled" trip. We arrived at a new park, exiting from SR 520 at the river's edge at 10 AM. Upon arrival, I saw a large tent near the entrance to the park. I was not there for a trip on the river. Read the full account of this event: Chapter X, Fruits of My Labor.

Premier Showing—Documentary on St. Johns River

On 3 December 2005, a sell-out crowd of 275 citizens gathered at the Brevard County Zoo for the premier showing of a PBS Documentary: *"River Into the New World: the St. Johns"*. In 2003, the St. Johns River Alliance contracted with Eagle Productions, based in Orlando, Florida to produce the film. On one occasion, using my boat, I transported the film crew staff along parts of the Upper Basin as they sought out the perfect shots of nature at work on the river. The film portrays the entire river's vast treasure of ecological, historic, cultural, economic and recreation assets. Premier showings were also held in the Middle Basin and the Lower Basin. The title of the documentary is very appropriate—the St. Johns River was the first river discovered in America.

The initial nationwide televised showing on PBS affiliate stations occurred in June 2006. Additional telecasts will occur over the next several years. The St. Johns River Alliance paid all costs for the documentary and will subsequently own the rights to the film. The alliance worked with the Florida Public Service and educational stations to air the film. Brevard County's local education channel has aired the film; future public viewings will be available. Indeed the river is receiving significant recognition for its true value as a majestic and historic river.

SJRWMD Purchases More Acreage in Upper Basin Marsh

On 2 February 2006, the editor of Florida Today newspaper reported: *"the SJRWMD completed the purchase of a significant parcel of 6,020 acres needed to complete the Upper Basin Restoration Project. The purchase from Fellsmere Farms included much-needed marshland located in northern Indian River County. The price was $63 million dollars which puts the total funding for this massive restoration project well over $200 million dollars"*.

The editor concluded the article by stating: *"Too often, government projects turn into Boondoggles that waste taxpayer money. That's not the case with the $200 million plus Upper Basin Restoration Project, or the Fellsmere lands just purchased. And the proof will come in the water flowing out of your tap in years to come"*. This writer concurs with the editor. Restoring the Upper Basin is going to be money well spent.

A Look to the Future

The future success of a healthy St. Johns River will rest with a new generation of individuals willing to challenge ill-advised development within the 25-year floodplain of the river. Local government has the responsibility to protect environmentally endangered lands. In Brevard County, citizens have voted twice over the past 15 years to continue taxing themselves so local government may continue purchasing environmentally endangered lands. Brevard County is fortunate to have three rivers; the St. Johns River (an American Heritage River); the Indian River and the Banana River (these two rivers are designated as a national estuary). The Atlantic Ocean and its beaches parallel the coast of Florida. Florida residents are indeed fortunate to live in such a natural paradise of beauty, tranquility and splendor to enjoy. There is a price for such an abundant natural environment. We must protect it.

I feel indebted to the St. Johns River and its wildlife, to share with the reader my experiences and commitment to protect this magnificent river and its many lakes. I attribute my sense of indebtedness to the thousands of hours I have enjoyed the river's resources; from camping and cooking breakfast on its banks to many great fishing trips with my family and friends. My hope is to inspire the reader to become personally involved in the protection of the river's rich history and natural beauty. You can ensure your children and grandchildren will enjoy one of nature's greatest gifts, the St. Johns River.

One person made a difference—I trust you will enjoy the book and that its contents will enlighten your understanding of the true meaning of Saving the St. Johns—an "American Heritage River".

Fishing Trip Memories: As promised, each chapter concludes with a brief story of a memorable fishing trip by this writer. 11 July 1961: I was wade-fishing for bass approximately 40 yards from the shoreline of Lake Winder on the St. Johns River in Brevard County. It was 30 minutes after sunrise on a beautiful morning. The water was quiet; no wind disturbed its glassy appearance. As I approached an old fence line visible at the water's edge, I felt a light tap on my black plastic worm. I quickly reeled up all the slack line as something began a steady, but deliberate move further into the open water of the lake. I set the hook—a big swirl flashed near the surface 30 yards away. Indeed, I knew I had hooked a very nice bass. However, the fish never jumped, but began running directly toward me, standing in 4 feet of water. I continued to reel in the slack line to catch up with its location, when all of a sudden, something struck the inside of my legs and kept right on going.

Before I could turn around, I caught a glance of a giant bass, finally jumping clear out of the water 10 yards behind me. I held on, even loosened the drag so as not to have a tiring fish break free as I was finally reeling in this trophy fish. I had landed a 10 pound, 8 ounce largemouth bass. It was the biggest bass of my long fishing career. That was 45 years ago. It is still the biggest bass this angler ever landed. I had this beauty mounted and have had the taxidermist touch it up a couple of time over the years. It remains on the wall, a pleasant memory of a day I will forever cherish.

CHAPTER II—

BEGINNING THE JOURNEY AND BEYOND

In the late summer of 1985, a major fish kill occurred in Lake Winder in Brevard County. It was not the first fish kill in the St. Johns River's Upper Basin, nor would it be the last. Lake Winder is located in a remote area, accessible only by boat. According to various government agencies and agricultural interests, numerous factors contributed to yet another disturbing loss to a great fishery. I hear the same excuses each time a fish kill occurs.

The agriculture community stated the fish kill was the result of significant storm events. Fish and Wildlife officials theorized heavy storms and hot weather, followed by several days of little or no sunshine needed to supply oxygen to the suffocating fish resulted in a tragic situation. Other state officials stated the event resulted from a combination of the above, but added that exotic hydrilla consumed significant amounts of oxygen.

My personal opinion led me to believe the major issue was the direct discharge of untreated runoff from thousands of acres of cattle ranches, citrus groves, row crops, and sod farms located upstream of Lake Poinsett. Poor water quality resulted in lowering the dissolved oxygen to a level below that needed to sustain the fishery. Ranch and farm operations encompass 90 percent of the land mass on both sides of the river. Agriculture farms used extensive dikes and canal systems with large pumps to offload excess untreated water into the river. However, permits for these pumps were routinely approved by state regulatory agencies.

On a visit to Lake Winder to inspect a hydrilla problem on the lake (unaware of the fish kill), my 150 horsepower outboard motor bogged down less than 100 yards into the lake. A thick mass of hydrilla overheated the engine. As I sat in my boat viewing the 2,000-acre "golf-course" of green, but decaying hydrilla, no water was visible, except for the small trail left by my churning boat propeller as it tried in vain to clear a path. I soon discovered that hydrilla was the least of the problem.

The stench of dead fish was overwhelming. In every direction, white objects were sticking up through the dense hydrilla. These white objects were dead fish; bass, bream, catfish, crappie, gar, and small bait-fish. Lake Winder was simply absent of oxygenated water. It was at this point this writer decided to raise public awareness of the many problems confronting the St. Johns River.

I contacted Bill Sargent, Outdoor Writer for Florida Today. Bill published an article announcing a public meeting to be held on 16 October 1985, at the Lone Cabbage Fish Camp on SR 520, west of Cocoa, on the St. Johns River. An estimated crowd of 75 concerned citizens attended. The massive fish kill was discussed. I explained my desire to form an organization to work with local and state agencies responsible for the river's health.

Representatives from nine different hunting, fishing, homeowner and business groups were in attendance. A decision was quickly reached to form an organization. I suggested our coalition be known as Sportsmen Against Violating the Environment (SAVE). After a lengthy exchange of ideas, SAVE was formed. Subsequently, the name "St. Johns River" was added to identify our specific mission. SAVE St. Johns River was founded.

At the original meeting, this writer was nominated to serve as President; I accepted the nomination. Mr. Ron Taylor, a member of the Florida Fly Rod club was nominated and agreed to serve as Vice-President. Other officers were installed at this initial meeting. One individual from each group present would represent their business or club position at future meetings. An opportunity to save the St. Johns River was underway.

SAVE grew to become a coalition of 30 organizations at the height of our many activities. Successful fund-raising events, especially bass tournaments were key to providing the needed funds to meet with various state agencies around the state. Over 100 local businesses supported SAVE with donated prizes given to both the anglers in the tournaments and to the public via drawings, etc. SAVE was in a financial position to challenge ill-advised development within the 25-year floodplain of the river.

My 30-year career in the aerospace industry and my eight-year service in the U.S. Air Force had taught me patience and discipline. At the time I did not understand how important both of these assets would be in the many issues SAVE would undertake.

First Official Action by SAVE

On 1 November 1985, I wrote a letter to various agencies and elected government officials. I introduced our new organization. I solicited their support in issues SAVE had identified from input within the newly formed group. At the time, we requested support for new policies and enforcement of existing regulations to return the river to as natural state as possible. I addressed the recent fish kills along the St. Johns River. Five issues SAVE identified were as follows:

1) Establish stringent controls for pumping agricultural runoff into the river with substantial fines for violations;
2) Clean up the hydrilla in the Upper St. Johns River;
3) Upgrade the state law concerning water quality standards to a meaningful level;
4) Purchase, or reclaim the marsh north of the Fellsmere Grade. Current pace is too slow;
5) Resolve the gross misinterpretation of the Marketable Records Title Records Act regarding sovereignty lands. Place titles to these lands back into hands of the state to be held in trust for the public.

Establishing Relationships with Government Agencies and Environmental/Conservation Groups

DEPARTMENT OF ENVIRONMENTAL REGULATIONS (DER)—As a result of my letter to the various government agencies, I was pleasantly surprised at the numerous responses I received. Some letters pointed me to the various agencies with responsibilities in the area of support I was seeking. Other letters offered phone numbers and names of personnel in their departments whom I should contact.

The Department of Environmental Regulation (DER) was especially helpful. I was asked to contact the Orlando office since their personnel worked more closely with issues SAVE wanted to discuss. Mr. Alex Alexander (deceased) and Jim Hulbert were most helpful in a meeting held in Mr. Alexander's office in early 1986. They responded to 20 written questions presented by SAVE regarding environmental protection for the St. Johns River. The meeting provided a better understanding of issues facing SAVE. An excellent partnership was established with the DER.

In late 1986, SAVE became involved in a key issue to get agriculture farms to retain the pumped discharge of polluted water in holding reservoirs on their property. We soon discovered that DER was addressing the issue, although SAVE felt the pace was to slow. We met with their officials and provided our input, requesting the matter receive expedited action.

In 1987, DER initiated a five-year consent order to all farms using pumps to discharge water into the St. Johns River. By 1992, every farm would be required to construct sufficient holding reservoirs to retain the first inch of stormwater on site. Every farm met the 1992 deadline, except the Duda Ranch. When DER indicated legal action would be taken, the Duda Ranch came into compliance. Subsequently, DER would change its name to Department of Environmental Protection (DEP).

SJRWMD—On 10 August 1987, I wrote a letter to Jim Swann, a member of the SJRWMD governing board. The letter resulted in a meeting with him at his Cocoa office. While he understood SAVEs desire to work with the SJRWMD, the meeting was not fruitful. SAVE sought assurance that the public would have use of the Upper Basin lands being purchased for restoration. I also wanted to discuss the restoration of the first two lakes in the Upper Basin, Lakes Hell N' Blazes and Sawgrass. I learned Mr. Swann was a developer who cared much about the environment; however, his generalized comments were of little help to us.

After learning what the Upper Basin Restoration Project encompassed, SAVEs relationship with the SJRWMD improved to one of mutual respect. I did not understand why the restoration effort excluded Lakes Hell N' Blazes and Sawgrass. I remain disturbed on this matter to this day. I met Maurice Sterling, SJRWMD Project Manager for the Upper Basin Restoration Project, who worked in partnership with the federal government's U.S. Army Corps of Engineers.

BREVARD COUNTY GOVERNMENT—SAVE worked with the Brevard County Commission in our efforts to obtain support on local and state conservation issues. Initially, in late 1986, I approached District IV commissioner Sue Schmitt regarding the need for a new public boat ramp at James Bourbeau County Park, located at the river off SR 520. The ramp and dock were in poor condition. I shared with her a drawing SAVE had prepared for a double launch ramp with a floating dock. Commissioner Schmitt obtained the funding for a new boat ramp, using the design drawings SAVE provided.

Within several months, the facility was built and continues to serve the public. In addition, lighting was installed to assist boaters when launching or removing their boats during darkness. The county dedicated the new facility to SAVE. An appropriate sign was installed. The sign remains in place.

SAVE designed double-floating boat ramp at James Bourbeau County Park off SR 520 on St. Johns River

I became acquainted with the county Parks and Recreation department during this effort. Jack Masson (now retired) was the Assistant Director. Jack assisted SAVE in the use of the county parks for fund-raising events.

In 1987, SAVE supported the county's purchase of approximately 1,100 acres of property from David A. Smith of Orlando. The property is located immediately south of SR 520 near the St. Johns River. The property was to be developed as a county regional park, bearing the name: F. Burton Smith, in honor of David's father (deceased). Working with county Commissioner Sue Schmitt, we understood the county was purchasing 1,100 acres of "land". Approximately 911acres of this property is located outside an illegal 18,000-foot dike in the open waters of Lake Poinsett.

Using taxpayer money from the Beach and Riverfront funds, the county paid approximately $1.3 million dollars for the property. Subsequently, the state of Florida determined the county had indeed paid for 911acres of sovereign submerged lands as part of the contract with David A. Smith, the "deeded" owner. In reality, the regional park contains approximately 189 acres, with a number of the acres in wetlands. Quite expensive "land" indeed.

Following the purchase of the 1,100-acre property by Brevard County, Mr. Smith approached the county seeking approval to develop a

720-acre site adjoining the regional park. The proposed project included a commercial center, over 800 residential units, and a golf-course. At this time in the process, SAVE had no knowledge the project included a golf course. SAVEs relationship with Commissioner Schmitt became somewhat difficult.

In 1989, SAVE voted to oppose David Smith's proposed Sabal Hammocks Project. SAVEs legal challenge was filed in June 1990. Mr. Smith had planned to deed the golf-course to Brevard County. In exchange, the county would assume responsibility for maintenance and operation of the dikes and pump station. SAVE called a special meeting with Commissioner Schmitt. We informed the commissioner of SAVEs decision to file a legal challenge against the proposed project. Commissioner Schmitt stated that if SAVE opposed the project, she would request the commission cancel its support for the golf course at the next commission meeting. As promised, the commission voted to remove the county's involvement in the proposed golf course.

In early 1993, SAVE requested the county commission write a letter to the state Department of Environmental Protection (DEP) stating Brevard County was opposed to development on state sovereign lands, except as provided in the Florida Statues. SAVEs request was based on the coalition's knowledge that a significant portion of the project was being developed on sovereign submerged lands; moreover, additional portions of the property was being built in the annual floodplain of the St. Johns River. The commission approved SAVEs request and forwarded the letter to the state on 13 March 1993. SAVE vigorously pursued the protection of sovereign lands.

In 1995, SAVE addressed the county commission regarding the development of the new city of Viera, south and west of Rockledge. SAVE requested the commission seek an Ordinary High Water Line (OHWL) determination of the 38,000-acre Duda Ranch. The commission supported SAVEs request and forwarded a letter to the state DEP requesting a cost estimate for the state to perform the OHWL survey.

The state responded, expressing their willingness to perform the survey and provided cost sharing funding requirements. Total costs for such an extensive survey would be $165,000. The state would pay 50 percent of the cost. The balance would need to be resolved between the county and Duda Ranch officials. As the reader will discover, the owner of Duda Ranch would sell over 14,000 acres of their land to the state, thereby avoiding an OHWL survey.

In 1997, a majority of the county commission provided SAVE a Resolution in support of our effort to win the federal designation of the St. Johns River as an American Heritage River. Commissioners Mark Cook,

Nancy Higgs and Truman Scarborough voted in favor of the resolution; Commissioners Helen Voltz and Randy O'Brien voted in opposition. SAVE desires to continue our successful working relationship with the county commission on future issues regarding what should be a common goal: protecting and restoring the St. Johns River.

FLORIDA FISH AND WILDLIFE CONSERVATION COMMISSION—A 1991 survey by U.S. Fish & Wildlife Service placed freshwater fishing at a $1.6 billion dollar industry in Florida. Anglers spending totaled over $600 million dollars. Almost $40 million dollars was generated in state sales tax, excluding fishing license sales. Every hour spent fishing was worth $5.31 to local economies. Over 2.7 million daily trips were spent fishing freshwater lakes and rivers. Approximately 33 percent of these trips were taken by visitors to the state. However, Florida has now lost its title as the "bass capital" of the world.

In late 1991, SAVE continued to conduct bass tournaments as a means of fund-raising. At this time, however, I reduced the number of bass that could be weighed in to 6 fish per team (two individuals). SAVE released all the bass back into the river. The idea caught on as other tournament sponsors were reducing the number of bass allowed at the weigh-in.

A continuing decline in the bass population was noted in the number of teams with a limit of 6 bass when compared to records from the 1984 thru 1990 limit of 14 bass per team. Typically, SAVE tournaments were drawing 85 to 110 boats.

The number of team limit catches in the earlier period (1984–1990) averaged 17 limits per tournament. During the 1991 thru 1997 tournaments, with a 6-fish limit per team, an average of 18 limit catches were weighed in. If we do the math, the average number of limit catches during 1984 thru 1990 was 238 bass. Using the 6-fish limit for 1991 thru 1997, the average number of limit catches was 108 bass. When comparing the two 7-year periods, limits of bass increased by 1, while the number of limit catches weighed in was reduced by over 50 percent. The St. Johns River was experiencing continuing fish kills which contributed to the demise of the bass fishery.

Many Florida residents are spending more money and time bass fishing outside the state. Texas and California are leading the nation in freshwater fishing revenue. The resolution to this problem is not only in restoration of Florida's lakes and rivers, but the need to further reduce the bag limit of bass per day to 3 per person with a slot-limit of 14 to 21 inch bass being returned to the lake immediately.

Florida fishery experts have disagreed with this writer on this issue since 1986; however the total bag limit was reduced from 10 to 5 bass per day several years ago. It is not enough. Ask a Texas Park Ranger or fishing guide. Fishing is excellent; the tourists dollars keep rolling in from out of state visitors every year. I speak from experience; Joyce and I spent our vacation fishing Lake Fork in Texas for four straight years, 1993 thru 1996. After viewing the video of our 1993 bass-catching trip, Bill and Wanda Rucker, of Cocoa, joined Joyce and me in 1994. Each one of us caught and released "over the limit" catches for 10 days. On each trip to the "bass factory" at Lake Fork, many vehicles bearing Florida license plates were well represented at the big fish camps on Lake Fork.

Joyce with a 5 pound bass caught in Lake Fork, Texas

On 11 October 1997, this writer provided the above data to a panel at a conference held in Palatka, Florida. The conference was jointly sponsored by the Florida Freshwater Game and Fish Commission and the Florida Bass Federation. Experts were in attendance from fisheries, DEP, COE, SJRWMD, South Florida Water Management District, University of Florida, and nationally known Bassmasters out of Montgomery, Alabama. When I spoke of our trips to Texas over the past four years, I got the attention of Dr. Jerry Shireman, Director of Fisheries for the Florida Freshwater Game and Fish Commission.

Dr. Shireman met with me for 30 minutes following the conference. I again emphasized slot limits (14 to 21 inch bass immediately released back into the water), bag reduction limits, and stocking programs where needed. I also suggested more catch and release lakes in Florida, where all bass must be released, such as is now the rule for the Stick Marsh near the Brevard and Indian River County line immediately south of the C-54 canal on County Road 507. Except for one or two small lakes where research is ongoing, I have no knowledge of any planned changes to fishery rules for the state of Florida.

Mr. David (Dave) Cox, a marine biologist (retired) documented numerous issues concerning the health of the St. Johns River. Dave made himself available to SAVE, from performing his favorite hobby, that of a playing a clown for the kids at SAVE fund-raising events, to appearing as an expert witness for SAVE in the legal challenge to the proposed Sabal Hammocks project.

In November 1998, Dave prepared a document entitled: *"Conceptual Plan for Restoration of the Upper St. Johns River—Lakes Hell 'n Blazes and Sawgrass and Protection of Lake Washington"*. The scientific data contained in the document clearly demonstrated the potential impact on the potable water supply from Lake Washington that serves 150,000 residents, primarily in the Melbourne area.

FLORIDA DEFENDERS OF THE ENVIRONMENT (FDE)—On 15 September 1992, after a lengthy discussion, SAVE voted to support FDE in the restoration of the Ocklawaha River, including the removal of Rodman Dam. SAVE forwarded a letter to the FDE the following day. SAVE became an alliance member group of FDE in 1993. The Ocklawaha River is a tributary of the St. Johns River, exiting into the river near Palatka.

FLORIDA WILDLIFE FEDERATION (FWF)—SAVE has been a long-time affiliate group of the FWF. The FWF supported SAVE through the years on our legal challenge to the Sabal Hammocks project. Manley Fuller, President of the FWF, won the support of its board of directors to join SAVEs request to intervene on behalf of the state following the state's decision to take over our lawsuit in 1995. However, SAVEs appeal to intervene on behalf of the state was turned down by the Appeals Court as was that of the FWF. The court stated the state had sufficient resources to represent the public interest.

SAVE Support Base

Many coalition member groups have supported SAVE with their time, talent and financial contributions over the past 20 years. I could never accurately list the 100 plus small businesses who donated prizes in support of our fundraisers, or the hundreds of anglers who financially contributed to SAVE. Also, there were over 160 dues paying citizens who simply wanted to receive SAVEs newsletter. I sincerely thank each of them and appreciate their support. The following list includes SAVE membership groups or affiliate organizations that have supported SAVE for many years:

1. Florida Wildlife Federation
2. Indian River Audubon Society
3. Classic Airboats
4. Neal's Tire Service
5. Hunt N Fish (Outdoor Shop)
6. Palm Bay Bassmasters
7. Florida Bass Federation
8. Patrick Air Force Base Bassmasters
9. STOPP—Stop Theft of Public Property
10. Quills Complete Anglers
11. Southland Bait & Tackle
12. Gheen Manufacturing
13. Mims Bait & Tackle
14. South Brevard Bass Rats
15. Forest Lakes Condo Assoc.
16. St. Johns River Valley Airboat Assoc.
17. American Ingenuity, Inc.
18. Christmas Riverboat Assoc.
19. Back to Basics Bass Club
20. Suncoast Bassmasters
21. Backcountry Fly Fishing Assoc.
22. Sunshine State Bassin Gals
23. Brevard County Air & Power Boat Squadron
24. Terri's Tackle Shop
25. Brevard County Bassin Belles
26. Toby "Ts" Worms
27. Ducks Unlimited (Brevard chapter)
28. Cocoa Bassmasters

29. Environet of Indian River County
30. M&S Welding
31. Florida Fly Fishing Assoc.
32. McDonald-Douglas Bass Club
33. Florida Wildlife Alert
34. Brevard County Sportsmen Assoc.
35. Gonzales Properties, Ocala
36. Space Coast Bassmasters
37. Lake Poinsett Homeowners Assoc.
38. Friends of the St. Johns (Sanford)
39. Lake Washington Homeowners Assoc.
40. Taylor's Lawn Service
41. Melbourne Bassmasters
42. Robert Wright Lawn Service
43. Spacecoast Bassmasters
44. Ole Fashion Airboat Rides
45. Bass Rods' Guide Service
46. SOS Satellite Service
47. Diversification Unlimited
48. Florida Fly Rod Club
49. Airboat Eco-tours

More Fishing Trip Memories—In 1963, my friend Bob Gonzales joined this writer on an exploratory fishing trip to a little known impoundment of fresh water pools created during the construction of the Kennedy Space Center complex on north Merritt Island, Florida. No boat ramp was available; we launched my 14-foot lightweight boat right off the nearest "open bank" area. This was truly a wilderness area; we never saw or came in contact with anyone during the trip.

I parked my pickup truck and returned to the boat. Bob was already seated. I was wearing a wading suit, with chest high protection to prevent any water from getting inside the suit. I leaned over and gently pushed the boat off the sloping bank. With a fishing rod in my hand, as I was about to push my weight off the bottom near the bank, there was no bottom. This writer sank over 5 feet before contacting the bottom; only my head was above water. My wading suit quickly began filling with water; I was desperate to get in the boat. Bob grabbed my rod, put it away and returned to help me into the boat. After a couple of minutes, I finally positioned my legs high enough into the air to drain the water from the suit. The two of us are serious anglers. Bob reacted as any angler would; **first things first—save the rod.**

We went fishing and had an enjoyable day, catching and releasing 15 nice bass in our 3-hour trip.

That same year, 1963 would provide yet another memorable fishing trip. I returned to Lake Winder where I had caught the giant bass a month earlier. This time I left my boat anchored in the same area of the lake. I wade fished along the shore, not realizing I had walked a considerable distance. It was another beautiful morning; I had already caught several nice bass up to 4 pounds. I decided to check my distance from the boat. I was easily 150 yards south of the boat. That was fine; however I had company in the water. A large alligator was trailing me. I quickly realized I had decided to keep the bass I caught for a photo of the catch before releasing them. I had them carefully attached to a thin chain stringer attached to my belt.

I made a conscience decision not to waste any time removing the bass from the chain stringer. I unclipped the stringer from my belt and let it slowly sink; meanwhile I headed very slowly toward the potential snake-infested shoreline marsh. It was an easy decision; the alligator had disappeared from the surface of the water. I thought the fish was what he wanted; I never saw the alligator again. I was not going to give up wade fishing; rowing a boat or drifting in the wind was not an option I would consider. Electric trolling motors had not been invented at the time.

CHAPTER III—

UPPER BASIN RESTORATION PROJECT

In my research for this chapter of the book, I was especially interested in how the state and federal agencies perform their role in environmental protection of natural resources along the St. Johns River. I located a great source of specific data I will share with the reader—Professional Paper SJ98-PP1, written by Maurice Sterling and Charles A. Padrea, both employed by the SJRWMD. I drew upon the professional paper, in part, for the historical account of events leading to the massive restoration project in the Upper Basin. As stated earlier, Mr. Sterling is the SJRWMD Project Director for the Upper Basin Restoration Project. The SJRWMD works in partnership with the U.S. Army Corps of Engineers of the Jacksonville District.

The subtitle of SJ98-PP1reads: *"The Environmental Transformation of a Public Flood Control Project"*. Indeed, in the early years flood control measures envisioned the draining of the Upper Basin marsh via multiple canals, dikes, and pump stations. Excess water would be offloaded through the Indian River Lagoon to the Atlantic Ocean. Construction of additional dikes and pump stations continued for years. Thousands of acres of the nutrient-rich soils of the floodplain were continually being opened for agricultural operations. Water management "improvements" served the agriculture community well.

In the early 1900s, the floodplain of the headwaters of the St. Johns River contained over 400,000 acres. In some areas, the shallow marsh of the floodplain was 30 miles in width. By 1970, over 70 percent of the floodplain had been converted to agriculture operations to support such enterprises as citrus, cattle, row crops and sod farms. With only 30 percent of the marsh still open, the hydrologic and ecologic characteristics of the headwaters were in peril. When major storms in the 1920s and 1940s devastated central and south Florida, the need for massive flood protection projects became very evident. For seven decades, there was no public outcry for action by the state or federal government to protect the natural resources of the floodplain.

In 1948, as a result of these major storm events, the U.S. Congress authorized the creation of the Central and Southern Florida Flood Control Project. The state of Florida created the Central and Southern Florida Flood Control "District" to partner with the federal government. The U. S. Congress and the state of Florida did not address the protection of the floodplain's natural resources.

Flood control protection of private property was the key interest of these new formed flood control districts. In fact, the Central and Southern Florida Flood Control District did not include the Upper Basin of the St. Johns River. Could it be the headwaters of the St. Johns River were considered already "lost" or that this "swamp" was not a part of the St. Johns River? At that time, they would have been 70 percent correct.

In 1954, however, the original congressional act of 1948 was amended to include the Upper Basin. In 1957, a project plan was prepared by the U. S. Army Corps of Engineers of Jacksonville. By 1962, the plan had been modified and adopted. The 1962 revision called for reduced flood stages in the Upper Basin by diverting large amounts of water during major storms to the Indian River Lagoon.

The Dawning of a New Day Slowly Leads to Restoration of the Upper Basin

The federal National Environmental Policy Act of 1969 required an Environmental Impact Statement be prepared for federally funded water projects. By 1970, the U. S. Army Corps of Engineers had initiated the Environmental Impact Statement for the Upper Basin Project. Soon, it was discovered that potentially serious environmental impacts would be encountered within the project. In 1972, the Upper Basin project was halted pending completion of a more comprehensive Environmental Impact Statement. The state of Florida determined that indeed the Upper Basin would suffer environmental degradation. In 1974, the state withdrew as a sponsor of the project. Environmental concerns included adverse impacts of large freshwater discharges to the Indian River Lagoon and the possibility for severe water quality and habitat degradation throughout the Upper Basin of the St. Johns River.

In 1969, the C-54 canal was operational. Major flood control benefits to the Upper Basin agricultural interests were realized. This process continued during the interim period whenever the Blue Cypress Lake subbasin water reached flood stage. Large volumes of stormwater discharges were diverted into the Indian River Lagoon. Documentation was incomplete on the consequences of stormwater diversion; however, current understanding is one of detrimental impact on the lagoon's fragile ecosystem. The commercial shellfish and sports-fishing industries were significantly impacted as well as the loss of sea grasses where bait-fish once thrived.

In 1972, the Florida Legislature created the SJRWMD. Mr. Dennis Auth served as the Executive Director. Following Mr. Auth's departure from the SJRWMD, he started an environmental consulting business. Subsequently, SAVE hired Mr. Auth to represent our coalition in the Sabal Hammocks legal challenge.

In 1977, local sponsorship for the Upper Basin project was transferred from the Central and Southern Florida Flood Control District to the SJRWMD. The transfer was established by Chapter 373, Florida Statutes. In 1979, the SJRWMD completed an extensive report that described the existing basin conditions. A Citizens Advisory Committee was created to work with the SJRWMD in designing a basic concept. In November 1980, the SJRWMD Governing Board adopted the concept.

In 1982, the U. S. Army Corps of Engineers agreed with the basic design criteria adopted by the SJRWMD as being economically justifiable and warranted federal participation. In February 1983, the Governing Board of the SJRWMD adopted a plan. The board requested the U. S. Army Corps of Engineers prepare a project plan and an Environmental Impact Statement.

The new plan, released in June 1985 was approved by the Chief of Engineers for the U.S. Army Corps of Engineers in August 1986. Local sponsor requirements were subsequently incorporated into the plan prior to the start of construction. These changes included a semi-structural project. The project would rely less on artificial controls and more on natural floodplains to store and conserve water.

Subsequent legal interpretations of the federal Water Resources Development Act of 1986 delayed the start of construction, resulting in a new funding arrangement for the project. The federal government would assume all costs for the construction and water conveyance elements. In May 1988, the Upper Basin Restoration Project was underway. At the time, it was estimated that a phased implementation of the project would be accomplished over the next 12 years; that would put completion of the project in mid 2000.

The original restoration project involved approximately 125,000 acres. The total acreage being restored has increased by 25,000 acres to

150,000 acres. In addition, the SJRWMD had to resort to imminent domain proceedings on two large parcels within the project's restoration area. These two events have contributed to the slowdown in completing the project.

Patience is Key to Success

The history of the early years in getting the Upper Basin Restoration Project underway provided this writer with a better understanding and increased my knowledge of the SJRWMDs management style. In late 1985, as I began to involve myself with their people, I received full cooperation from most every person I encountered. I will address two situations where SAVE had some "difference of opinion" in working some issues with the SJRWMD. One situation involved the Proposed Sabal Hammocks Project (refer to Chapter V). The other situation is documented in Chapter IV, Restoring Lakes Hell N' Blazes and Sawgrass.

In late 1985 and early 1986, SAVE was seeking to determine who the "players" were in order to have a voice in restoring the Upper Basin of the St. Johns River. The restoration project was a combination of state and federal agencies. I wrote letters to a local member of the SJRWMD Governing Board; the Department of Environmental Regulations; the Florida Freshwater Game and Fish Commission; selected members of the local legislative delegation; the Florida Governor; the local U.S. Representative in congress; and to Florida's two U.S. Senators.

This writer received supportive responses directing a path to the proper departments. U. S. Senator Lawton Chiles responded and followed up with this writer, by letter twice more during a 6-week period. Senator Chiles would later serve Florida citizens as Governor. He was a great leader and performed his duties in the interest of the "people" of Florida.

The Florida Freshwater Game and Fish Commission and the SJRWMD provided several airboat tours for SAVE representatives to view many areas of the Upper Basin Restoration Project. I observed earthen plugs being placed into agricultural canals for the purpose of forcing polluted canal waters back into the marsh. The water would "sheet flow" over the marsh, thereby removing many pollutants before reaching the river. In addition, large and small control structures were installed for storage and control of excess water. This water would be released into the marsh during the "dry season". This process would remove contaminants and provide a more natural flow of water on its journey northward into the St. Johns River.

Environmental protection is now being achieved, although the total project has yet to be completed. The natural hydrologic function of the Upper Basin Restoration Project creates near natural cycles of the once pristine headwaters of the river. Soil and vegetation characteristics are being maintained. Other benefits include improved water quality, which has enhanced fish and wildlife habitat.

Flood protection measures necessitated the Upper Basin Restoration Project include over 100 miles of levees, six large-capacity gated spillway structures, and 16 small water control structures, culverts, and weirs. As stated earlier, the project totals 150,000 acres. During flood conditions, the project may contain over 550,000 acre-feet of water—that could cover 86 square miles of water at a depth of 10 feet. The following paragraphs address the specific means of operating and controlling water storage areas to ensure proper water levels are maintained in these "reservoirs" to offset any potential for major flooding.

Water Storage Areas in the Upper Basin

Perhaps the reader is aware of the Marsh Conservation Areas (MCAs) and the Water Management Areas (WMAs). Do you know their specific purpose? The MCAs have major environmental features. They contain existing and restored marshes. The MCAs provide temporary storage of floodwaters generated from adjacent upland areas. The MCAs reduce the amount of potentially damaging quantities of fresh water into the Indian River Lagoon. The MCAs total over 16,000 acres.

The WMAs are located south of the Fellsmere Grade and east of the Blue Cypress MCA. These areas include the Blue Cypress WMA and the St. Johns WMA. The Sawgrass Lake WMA is located south of US 192 and east of the St. Johns MCA. These MCAs are located in the river valley; consequently, they are deeper reservoirs, owing to the soil subsidence from years of agricultural activities. The WMAs will provide long-term water supply and temporary flood storage of agricultural pumping operations and gravity flow from the east. The WMAs improve water quality by separating agricultural discharge from the better quality water in the St. Johns River marsh.

Entrance to Blue Cypress Recreation Area off SR 512 south of Fellsmere, Indian River County

Marsh Floodplain Destruction Approaches Wipeout

In 1980, it is worth noting that only 27,000 acres of marsh existed between the Fellsmere Grade in north Indian River County and south Brevard's Lake Washington. This area represented less than 20 percent of the historic floodplain. An additional 36,000 acres of restored floodplain between these two referenced points are being reclaimed.

Lakes Hell N' Blazes and Sawgrass are located south of US 192, in the restoration area, but were not included in the Upper Basin Restoration Project. I mention the fact here because this chapter is dedicated to restoring the Upper Basin. SAVE has supported the SJRWMD since 1988 to secure funding to restore the two lakes. In my opinion, from the beginning, these "dead lakes" should have been included in restoration of the Upper Basin.

The demise of these lakes is stated in the following chapter. This writer will continue to explore all options to restore these lakes, from local government, state legislators, and the federal government. A "true" Upper Basin Restoration Project will not be complete until Lakes Hell N' Blazes and Sawgrass are restored.

Another project in work by the SJRWMD that will benefit the St.

Johns River as well as the Indian River Lagoon is known as the C-1Rediversion Project. Funding problems at the federal level are slipping away (perhaps due to Hurricanes Katrina and Rita). The SJRWMD plans to provide the balance of funding, estimated to be $7 to $8 million dollars to get the project underway in 2008. The project will divert fresh water via the C-54 canal into the Lake Sawgrass Water Management Area. This project will return additional historic water flow back into the St. Johns River. SAVE supports the C-1 Rediversion Project.

A public/private partnership with Ducks Unlimited provided the funding for a waterfowl management area within the C-54 management area. The project is located west of the nationally known "Stick Marsh", a sportsman's haven for largemouth bass fishing. The waterfowl project is maintained on-site by the Florida Fish and Wildlife Conservation Commission. This approach should ensure a productive waterfowl area as part of the Water Management Area system.

Upper Basin Restoration Project—an Expensive Undertaking

The total cost for the Upper Basin Restoration Project will exceed $200 million dollars. In late 1997, actual expenditures were at $172,600 dollars. The partnership between the SJRWMD and the federal government has been a good partnership. The federal government provides the cost of engineering design and construction. The SJRWMD is responsible for land acquisition costs. A 50/50 split of cost is shared regarding development of recreational facilities. Using the 1997 data, cost sharing has worked out to be a near split of the $172,600. Engineering design and construction is almost equal to land acquisition.

Several factors came into play in the SJRWMDs public acquisition of environmentally sensitive lands. The land acquisition program, for the most part was favorable with landowners. Factors included:

1) Successful SJRWMD bond efforts for land acquisition;
2) Increase in documentary stamps for Save Our Rivers program;
3) Positive public sentiments toward programs to improve the environmental conditions;
4) Willingness to sell by landowners rather than meet more stringent development criteria;
5) Economic recession marked by high interest rates and rising land values.

As previously stated, in two instances the SJRWMD exercised its eminent domain powers to acquire two large parcels needed for the restoration project.

Public Will Benefit from Conservation Areas Throughout the River's Length

The Upper Basin Conservation Areas described in the following paragraphs provide great opportunities for the public to visit. Some are primitive; they were planned to be so. Others are more open and easier to explore. Camping is allowed in many of the conservation areas. The following paragraphs provide the reader with useful information on the many recreational opportunities available in the Upper Basin Conservation Areas.

Three Forks Marsh Conservation Area—The area contains 52,000 acres. It is part of the Upper Basin Restoration Project. The first discernible channels (or streams) of the St. Johns River arise within the marsh south of Lake Hell N' Blazes. The original floodplain communities of this area were severely impacted by diking and draining for agricultural uses. Water management areas now separate and improve the quality of agricultural water before the water enters the river.

Restoration will greatly improve the habitat for waterfowl, wading birds, river otters and the fishery. A boat ramp has been in place for several years. Hiking, biking and bank fishing on the levees will be possible. In addition, seasonal hunting will be allowed. Primitive camping at designated sites, air-boating, boating, fishing, canoeing and wildlife viewing are some of the opportunities that will occur upon completion of this in-work restoration project.

The Three Forks Conservation Area will contain an 8,000-acre lake when the restoration project is completed. The area was named: "Thomas O. Lawton Recreation Area" in honor of a friend whom everyone called Tom. The lake naturally became known as Lake Lawton. Tom had many friends; he was a special friend to the St. Johns River. Many of us were present at the dedication ceremony. He was humbled by the honor. Tom strongly supported SAVE for many years.

I served with Tom on the Southern Recreational Advisory Committee for a number of years. This SJRWMD sponsored group provided input to encourage the SJRWMD to support the public's desire for recreational opportunities in the Upper Basin of the St. Johns River. I speak for thousands of citizens who look forward to good fishing, hunting and wildlife viewing opportunities at the Thomas O. Lawton Recreation Area. This area will indeed be very popular for the outdoor enthusiasts. Tom would have been pleased to see the lake fill up. However, at age 87, Tom passed away. His legacy will continue. Future generations will benefit from Tom's dedicated work in restoring the Upper Basin of the St. Johns River.

Blue Cypress Conservation Area—The area contains over 54,000 acres. It is part of the Upper Basin Restoration Project. It is located between SR 60 and the Indian River/Brevard County line at Fellsmere Grade. The world famous "Stick Marsh and Farm 13", located at the Indian River/Brevard County line, contains 6,000 acres of excellent fishing and wildlife viewing. The property was previously owned by agriculture interests. The reservoir was created to prevent agricultural discharges from entering the St. Johns River marsh.

An approximate 3,000-acre orange grove previously occupied the "Stick Marsh" portion of the reservoir. The name suits the description of the old orange grove. The orange grove was flooded, along with some other hardwood trees. Previously, Farm 13, a vegetable farm occupied the remaining 3,000 acres. The two areas are connected by a center canal area that is open to boat navigation. I have enjoyed many trips to this anglers' paradise.

On one fishing trip to the "Stick Marsh", Joyce and I caught and released about 30 bass up to 6 pounds during a 2-hour stop at a spot where we observed bass feeding on baitfish. We were video taping our success until we both had fish on at the same time. Two men fishing nearby got our attention. They said to us: *"you two catch the fish; we will get you on our video camera"*. As we scanned the immediate area, there were two more boats sitting idle 30 yards away watching the two of us. It was one of those times Joyce and I will never forget; more so because these other anglers quit fishing to watch us catch many bass in a very short time period.

On other trips to the Farm 13 side of the reservoir, we were successful in catching plenty of Crappie (called "specks" in Florida). Many times on successive casting of small artificial jigs using our ultra-light spinning rods, we would soon have all the "specks" we wanted to clean. The daily bag limit is 25 "specks", however, the Florida Fish and Wildlife Conservation Commission rules require the immediate release of any bass caught. I agree with this rule; the results of "catch and release" provide most everyone an enjoyable day of bass fishing.

Nearby Kenansville Lake is another great fishery and hiking area. Small boats are used to launch from a ramp on a levee. The lake is accessible via a canal alongside the levee. Bank fishing from the levee is allowed. The lake is approximately 9 miles east from US Highway 441on a dirt road at Kenansville.

View of nationally known "Stick Marsh" in north Indian River County—a trophy "catch and release" bass fishing paradise

Blue Cypress Lake, one of the pristine lakes in Florida is located off SR 60, southwest of the Stick Marsh. A private fish camp with lodging is located at the boat ramp. The east side of the lake in lined with large cypress trees. I have experienced great fishing on Blue Cypress Lake. I am always amazed at its beauty, in particular the huge cypress trees.

There is excellent wildlife viewing from many of the levees surrounding this vast conservation area. Typical wildlife includes great blue herons, white ibis, snowy egrets, limpkins, wood storks, ospreys and bald eagles. On occasion, you may spot a deer in the shallows or upland areas. Levees are open to bicycling, hiking, canoeing, air boating, seasonal hunting, and primitive camping in designated areas.

The primary function of this conservation area is to reduce flooding, restore and maintain natural hydrologic cycles and protect water quality. Reduced offloading of freshwater into the Indian River Lagoon is an environmental benefit.

Bull Creek Wildlife Management Area—This area contains 23,470 acres. The property was acquired by the SJRWMD for flood control as part of the Upper Basin Restoration Project. The area provides flood protection and recreational opportunities. The main entrance to this area is located 22 miles west of the I-95/U.S. 192 interchange. From U.S. 192, turn left onto Crabgrass Road; the road dead-ends at the entrance, approximately 6 miles from U.S. 192. A brochure is available for use in an 8.6-mile self-guided tour on the loop drive. The Florida Trail Association maintains the portion of the Florida National Scenic Trail that passes through the property. Friends of Bull Creek, a citizen volunteer group, helps with cleanup, trail maintenance and restoration planting.

This area offers seasonal hunting, fishing, hiking, bicycling, horseback riding, camping at designated sites, canoeing and wildlife viewing. Levee 73 is open for picnicking, hiking and bicycling. Visit Bull Creek and enjoy this wilderness of undisturbed natural beauty.

Fort Drum Marsh Conservation Area—This property contains 20,862 acres. It is located in the southwest corner of Indian River County, between SR 60 and the Florida Turnpike, twenty miles west of Vero Beach. To access this area from I-95, exit at SR 60, travel west to "Twenty Mile Bend". The marsh area represents the southern most reach of the St. Johns River headwaters. The area was acquired as part of the Upper Basin Restoration Project. The natural communities include dry prairie, pine flatwoods, hardwood swamp and freshwater marsh.

Recreational opportunities include hiking, horseback riding, fishing, bicycling, primitive camping at designated sites, canoeing, environmental education and seasonal hunting. Wildlife viewing provides a look at sandhill cranes, wood storks, caracara, bald eagles, deer, turkey, and a large population of feral (wild) hogs. A boardwalk provides an opportunity to hike through a hardwood swamp to reach Hog Island where trails and primitive camping is available. A picnic pavilion and tables are located adjacent to Horseshoe Lake. The Florida Trails Association developed the trails and primitive campsites. Horseback riding is not allowed on the levees.

River Lakes Conservation Area—This area consists of 36,156 acres. Over 14,000 acres of this property is very special to this writer (refer to Chapter X). Located north of the Upper Basin Restoration Project which ends at U.S. 192, this conservation area includes Lakes Washington, Winder and Poinsett. Access to Lake Washington is via a county park boat ramp at the end of Lake Washington Road. A large pavilion with picnic tables and bathroom facilities are located in this park. Access to Lake Poinsett and Lake Winder is provided at Lake Poinsett Lodge, James Bourbeau county park and Leroy Wright Recreation Area, all located off SR 520 west of Cocoa.

A fixed crest-weir (dam) at the north end of Lake Washington prohibits most boats from proceeding north on the river, except when the water levels overflow the weir. An airboat ramp is available at the weir for travel north toward Lake Winder. The weir was put in place to ensure proper water level is maintained in Lake Washington, as this lake provides the potable water supply for over 150,000 residents in the Melbourne area. Minimum flows north are provided at the weir to maintain some degree of water flow into the river north of the weir. Lake Winder is the next lake north of Lake Washington.

The River Lakes Conservation Area provides important habitat for fish and wildlife, including several threatened or endangered species. These include the bald eagle, sandhill cranes, wood storks, and river otters. I have observed a large population of the once threatened alligators. Special hunt permits are now issued to keep their numbers in check. The diversity of fish and wildlife provides an important recreational resource for anglers, hunters, boaters and bird-watching.

Several airboat tour services are available in the River Lakes Conservation Area. My personal choice for an exciting, but educational tour of the area's wildlife is Airboat Eco-Tours. Capt Rick Thrift, owner of Airboat Eco-Tours, is a licensed captain. Your safety on the water is first priority for him. Captain Rick provides ample time for passengers to ask questions.

Capt. Rick Thrift with a tour group leaving Leroy Wright Recreation Area at SR 520 on the St. Johns River

Captain Rick has an excellent knowledge of the plants and wildlife in the River Lakes Conservation Area. He is a true conservationist willing to share his commitment for a clean river and treats his clients to an enjoyable trip on the river. Airboat Eco-Tours operates at Lake Poinsett Lodge, off

SR520, just west of I-95, at the end of Lake Poinsett Road. For information or reservations, Captain Rick can be contacted at (321) 631-2990, or www.airboatecotours.com.

Triple N Ranch Wildlife Management Area—The area consists of 15,391 acres. This area is located in Osceola County, south of U.S. 192 and east of U.S. 441. This property provides an upland buffer to the waters of Crabgrass Creek, Jane Green Creek and the Upper St. Johns River Basin. Flatwoods, dry prairie, oak scrub, cypress domes and mixed broadleaf hardwood swamps are found in the area. This property was acquired by the SJRWMD and the Florida Fish and Wildlife Conservation Commission to protect wildlife habitat and water resources. Access to the area: from Melbourne, take US 192 at the I-95 interchange; proceed west for 30 miles; entrance is 3.5 miles east of Holopaw on the south side of U.S. 192.

Significant wildlife species abound in the area, such as the sandhill crane, bobcat, river otter, deer and turkey. Seasonal hunting is allowed. Bicycling, wildlife viewing, horseback riding, hiking, primitive camping at designated sites and nature study are available. The area is open from sunrise until sunset. The Florida Fish and Wildlife Conservation Commission staff manages the property.

Other Conservation/Recreation Area Opportunities Available for PublicUse—In the northern (lower basin area), 21 conservation type outdoor adventures are available; from the smallest site (274 acres) at Stokes Landing, located north of St. Augustine, to the largest site (27,333 acres) at Lochloosa Wildlife Conservation Area, located southeast of Gainesville.

In the central (middle basin area), 13 conservation type outdoor adventures are available; from the smallest site (2,352 acres) at Lake Norris Conservation Area, located in eastern Lake County, to the largest site (29,145 acres) at Seminole Ranch, located at the convergence of Orange, Brevard, Volusia and Seminole counties between SR 46 and SR 50, east of Orlando.

Other SJRWMD lands are located in the southern (upper basin area), but are not a part of the St. Johns River Upper Basin Project area. For more detailed information on all basin areas, obtain a copy of the *"Recreation Guide to District Lands"*. The guide provides detailed information on the northern, central and southern regional areas where public usage of lands is allowed. Contact the SJRWMD Office of Communications and Governmental Affairs, 4049 Reid Street, Palatka, Florida 32177. You may also contact the office on the internet at sjrwmd.com.

In this chapter of the book, I included several of my personal experiences enjoying the SJRWMD lands. Joyce and I have visited other sites in the Upper Basin of the St. Johns River. I trust the information provided will encourage the reader to visit these special places. These lands are a significant public investment that will serve the citizens of Florida for many years. We plan to visit other recreation and conservation areas in the Middle and Lower basins. There is something for everyone to enjoy.

More Fishing Trip Memories—In 1970, my 6-year old son, Robert (Robbie) accompanied Dad on an early morning fishing trip to Middle River, a part of the St. Johns, lying between Lakes Winder and Poinsett. We began fishing for speckled perch in shallow water; these tasty fish were busy preparing their spawning beds. After casting my ultra-light spinning rod near several weed bed areas for approximately 15 minutes, I realized Robbie was no longer fishing. I turned to discover he had curled up under the console of the boat and fell asleep. I did not disturb him; after all, it was our time on the river together.

I began to catch a few of the "specs". Fish flopping around in the boat must have awakened him. He jumped to his feet and mumbled a few words. He said he was waiting to see if the fish were going to bite. He then joined in the fun; together we caught 10 fish. The number of fish was not important; this Dad wanted to instill the love of fishing in his kid. Robbie seemed to enjoy fishing, but the "sleeping scene" would be repeated several more times during the spawning season of the speckled perch.

On a subsequent trip in the early summer, Robbie and I returned to Middle River. His first bass fishing trip with Dad would provide a lifetime memory for both of us. Using a small Zebco 202 rod and reel, and casting a black plastic worm near the water's shoreline, Robbie screamed: "I got a big fish, help me Dad". The fish was taking line and moving downstream in the slow current of the river. Robbie asked again "Here Dad, you get it in, it's getting away". I gripped his hands onto the reel and said: "Robbie, you can bring it in, it won't break your line; keep winding the reel".

When the fish jumped out of the water, I realized he had a very nice bass on and I was not going to participate in handling the fish until Robbie could get the fish to the boat. He landed his first and biggest bass (he is now 42 years old). I had a camera onboard and took a photo of him holding his fish right at the spot he caught it. That photo is framed and displayed in our home. Since it was his first bass and he was only 6 years old, I had the 5-pound bass mounted; that bass is displayed on the wall of his home. Again, a wonderful memory for both of us. By the way, Robbie never went to sleep in the boat again. He now owns a bass boat and is frequently on the St. Johns River.

CHAPTER IV—

RESTORING LAKE HELL N' BLAZES & SAWGRASS

Lakes Hell N' Blazes and Sawgrass were the best kept secrets among anglers until the late 1950s. Lake Hell N' Blazes, the first lake on the St. Johns River receives its water from an area known as "Three Forks". Three small streams merge into a larger stream before entering the south end of the lake. The lake's water supply is supplemented by runoff from other parts of the upper basin floodplain. Lake Hell N' Blazes northerly outflow constitutes the beginning of the St. Johns River. The river continues a meandering and lengthy journey of 310 miles, exiting into the Atlantic Ocean at Mayport, north of Jacksonville.

Camp Holly Fish Camp, off US 192 west of Melbourne—Lake Hell N' Blazes and Sawgrass are located south of the camp

Lake Sawgrass is the second lake in a river of lakes located north of Lake Hell N' Blazes, and south of US Highway 192 in Brevard County. The lake's name is appropriate, being surrounded on the north, west, and south with a very hearty blanket of sawgrass.

Several legends exists as to how Lake Hell N' Blazes received its name. This writer prefers to believe the story of two friends fishing on the lake. At the time, numerous floating islands adorned the lake and were constantly changing their location. Windy days would cause the floating islands to move around depending on wind direction. These friends had traveled by boat from Camp Holly, located immediately south of U.S. Highway 192, west of I-95.

After a successful fishing trip and paying little attention to wind conditions, they decided to head back to the launch ramp. Although it was a hot summer day, there was a good breeze on the water. As they departed, the shoreline scenery had changed. Their return route to the river was almost blocked by several small "floating islands". One friend said to the other: *"I see the channel of the river, it's over there"*. After traveling a short distance, their conceived river channel disappeared into the marsh. They had entered a small cove. His partner responded: *"where in the hell n' blazes are we anyway"?* The name was born.

Lake Hell N' Blazes encompasses 260 acres. Lake Sawgrass, somewhat larger, contains 475 acres. Both of these lakes were among the best fisheries in the state of Florida for many years. I fished the lakes in the 1960s with much success. Sadly, by the early 1970s, I witnessed the slow, but serious degradation of the water quality in both lakes. Fish and wildlife habitat had been significantly impacted.

Demise of a Great Fishery

Agricultural interests owned most of the marsh and floodplain in the Upper Basin of the river. Dredging and diking operations were being attempted by the 1920s. Subsequently, in order to protect expanding agriculture operations, two major canals; the C-40 on the east side of the marsh and the South Mormon Outside Canal on the west side, were constructed.

During storm events, stormwater rushes down these canals, offloading all types of rotting organic matter into Lake Hell N' Blazes. By the early 1980s, over 70 percent of the original floodplain had been lost to agricultural operations. In the early 1900s, the marsh was 10 to 30 miles wide in many areas. By the 1970s, it had been reduced to approximately 3 miles wide near Lake Hell N' Blazes.

The great fishery in Lake Hell N' Blazes was destroyed by years of discharging polluted water from nearby farms into the marshy headwaters of the river. Major storm events necessitated the offloading of excess water from the farms in the Upper Basin. The presence of wildlife in the area has been seriously disturbed.

The natural beauty and splendor of this once peaceful and pristine paradise, while degraded, can still be enjoyed, except for the absence of the fishery. No roads, residential subdivisions, wires hanging from telephone poles, or any other hint of development exits in the Upper Basin headwaters south of U.S. Highway 192 until one reaches the small community of Fellsmere. Airboat tours of the marsh areas near Lakes Hell N' Blazes, Sawgrass and Washington are available at Camp Holly.

During high water levels in the area, fish migrate from Lake Washington to Lakes Hell N' Blazes and Sawgrass. However, poor water quality in these lakes forces the fish back into Lake Washington. Data from the Florida Fish and Wildlife Conservation Commission, hereinafter referred to as FWC, indicates their ventures to these lakes are short in duration. Florida's prized largemouth bass no longer spawn in Lake Hell N' Blazes.

In my 30-year career in Aerospace, preceded by 8 years of service in the U.S. Air Force, I acquired a keen sense of tolerance and respect for the opinion of others. Over these 38 years, I was part of a great team of dedicated professionals. For the past 20 years, this writer has served as President of a nonprofit corporation in many areas of responsibility. My tolerance was truly tested, but not my respect for differing opinions.

In 1988, SAVE began working with the FWC and the SJRWMD, seeking support for a feasibility study on Lakes Hell N' Blazes and Sawgrass. The purpose of the study would be to determine the feasibility of restoring the two lakes. In this chapter, the reader will discover how SAVE approached local, state, and federal agencies, including the U.S. Army Corps of Engineers to form a partnership in restoring the first two lakes on the St. Johns River. As the complexity of restoration unfolded, SAVE authorized this writer to represent our coalition in future meetings, site visits, travel, etc.

Also in 1988, the SJRWMD approved SAVE as a member coalition of the Upper Basin Recreation Advisory Committee. I represented SAVE at many subsequent meetings over the following 10 years. At one of these early meeting, I recommended this committee support the feasibility study to restore Lakes Hell N' Blazes and Sawgrass. The motion was approved; the message was sent forward to the SJRWMD staff. I became personally involved with the SJRWMD and the FWC to restore these "dead" lakes. I am not a biologist or a hydrologist, just a river advocate committed to a saving a great river.

Lake Washington is located approximately six miles north of Lake Hell N‘ Blazes. As a potable water supply, Lake Washington is designated as a Class III water body. The health and well-being of thousands of citizens depend upon the water intake structures and treatment plant to remove pollutants detrimental to public health. Today, according to a recent article published in Florida Today, the 4-year old water treatment plant is experiencing an increase in sediment levels at the plant. The question arises: How much longer can the plant treat the ever-increasing amounts of pollutants being transferred from Lakes Hell N' Blazes and Sawgrass into Lake Washington?

In the spring of 1992, Lake Washington experienced it's first reported fish kill. High water levels existed throughout the Upper Basin of the river. The FWC documented over 13,000 fish suffocated due to low amounts of dissolved oxygen. High winds triggered the uprooting of sediments. Lakes Hell N' Blazes and Sawgrass contained massive amounts of hydrilla and water hyacinths. Rising water levels increased the current flow from these two lakes northward toward the U.S. Highway 192 Bridge at Camp Holly. The bridge acted as a dam as tons of the sediments, hydrilla and water hyacinths became entangled against the pilings of the bridge. Within a few hours, the mass was solidly backed up river-wide, bank-to-bank.

Lake Washington lies approximately one mile north of U.S. Highway 192. The structural integrity of the bridge was threatened as winds and current continued to compact tons of the floating river-clogging "garbage" against the bridge. Equipment was brought from south Florida to break up the massive jam. Trucks hauled off much of the hydrilla and water hyacinths; however, organic matter (sediments) and a significant amount of hydrilla broke free and drifted the remaining mile into Lake Washington. Several years later, a new high span bridge replaced the one that had incurred potential structural damage.

In late 1992, an Upper St. Johns Fishery Study Team of the FWC began a field study of the sediment accumulation on the bottom of Lakes Hell N' Blazes and Sawgrass. Technical details of that study may be confusing to the reader; therefore this writer will proceed to the findings reached by the study team. The combined sediment volume for both lakes revealed the need for sediment removal in order to protect Lake Washington's potable water supply. The study concluded that restoration of these lakes, in addition to protecting the potable water supply of Lake Washington, would restore the fishery in Lakes Hell N' Blazes and Sawgrass, and improve wildlife habitat. The report was forwarded to the SJRWMD.

SJRWMD Contracts for Study

In 1994, the SJRWMD acted on the technical report provided by the FWC. The agency contracted with the Institute of Food and Agricultural Sciences at the University of Florida to undertake a study to determine the rate at which sedimentation was occurring. The report stated that sedimentation rates were determined to be high during flood events and/or physical disturbance, such as major canal construction.

The report also stated that sedimentation appears to have stabilized or even slowed in Lakes Hell N' Blazes and Sawgrass over the past 10 to 15 years, while sedimentation has steadily increased in the north end of Lake Washington. This confirmed the fact that Lakes Hell N' Blazes and Sawgrass are at near capacity for storing sediments. More sediment is being transported into Lake Washington. This is not good news for the public water supply at Lake Washington.

The findings of this scientific study should sound an alarm regarding the public water supply coming out of Lake Washington. The situation leaves this writer with two concerns. First, as sediment levels increase, future water treatment cost at the plant will likely increase. My second concern; Lake Washington could experience a potential catastrophic failure from raging hurricanes. Can the present weir withstand such an assault? Potentially, this 4,000-acre lake may require very expensive restoration.

My point is simple: hurricanes and major storms will strike the Upper Basin again and again. Why should the state risk the potential for negative public health issues and/or added expense to treat an ever-present problem? Why should the sad environmental condition of Lakes Hell N' Blazes and Sawgrass continue to threaten the water quality of Lake Washington?

The historic St. Johns River received a federal designation as an American Heritage River in 1998. Following the recent $63 million dollar purchase of 6,000 acres from Fellsmere Farms, over $230 million dollars have been spent to restore the Upper Basin. I have witnessed the benefits and fully support the Upper Basin Restoration Project. During the 1980s, why were these lakes not included in the Upper Basin Restoration Project? This missed opportunity has passed. Other attempts have failed; however, the users of the potable water supply should demand restoration of Lakes Hell N' Blazes and Sawgrass.

Attempts at Restoration Continue Drifting to Dead End

In recent years, the FWC and SJRWMD made an effort to secure funding to restore these two lakes; however, costs estimates continued to escalate. My opinion: part of the issue could have been that the SJRWMD viewed these lakes as a fishery; therefore the FWC should have taken the lead action. The FWC may have considered the issue to be an environmental and water quality issue; therefore they may have been waiting upon the SJRWMD to take the lead. In any event, the result was: too little action and too late in recognizing who would lead—just a river advocate's thinking. Both agencies may have done their best; documentation exists to demonstrate the effort was pursued.

In a letter dated 23 September 1996 to Mr. Eric Bush, Department of Environmental Regulation (DER) from Mr. Maurice Sterling, SJRWMD Director for Division of Project Planning, Mr. Sterling sought the COEs review of a conceptual restoration project published by the FWC. Mr. Sterling requested the COE provide the SJRWMD with recommendation in pursuing permitting for the project, assuming the FWC would partner with the SJRWMD to move forward on the project.

In my attempt to maintain a chronological order of events, I must inject an issue regarding access to these two lakes should they at some point receive the needed restoration. As stated earlier, the only access available to Lakes Hell N' Blazes and Sawgrass is from U.S. Highway 192 at the bridge over the river. On or about 4 December 1997, I attended a meeting of the Florida Department of Transportation (DOT), held in the city of Rockledge. DOT scheduled the meeting to inform the public about replacing the old bridge on U.S. 192, with a new higher span bridge that would have a 12-foot clearance at normal water level of the river.

At the meeting, U.S. Representative Dave Weldon's aide, Mr. Chase, stated the new bridge did not require a higher span elevation. Mr. Chase further stated that Rep. Weldon said it would be a waste of tax payer money to construct the new bridge 12 feet above the water level of the St. Johns River. Mr. Chase stated the river was not navigable south of the bridge; therefore there was no need to expend an extra $2 million dollars to elevate the new bridge.

Following Mr. Chase's appearance, I addressed the DOT staff. I assured the staff that indeed the river was navigable south of the bridge. I publicly invited Mr. Chase to contact Rep. Weldon and allow this writer to transport both of them on a trip to Lake Hell N' Blazes, approximately five miles south of the U.S. Highway 192 bridge on my 20-foot boat.

Florida Today published an article on 5 December 1997, accurately documenting the meeting. Mr. Chase never accepted my offer. The high span bridge was constructed to include the 12-foot clearance from the water level of the river. Boats, including airboats, can now navigate the Lake Washington to Hell N' Blazes route via the new bridge span, even during high water levels.

On 31 March 1998, Mr. Sterling wrote another letter to the COEs Mr. Richard Bonner, Deputy District Engineer for Project Management. Mr. Sterling informed Mr. Bonner that the SJRWMD and FWC wanted to proceed with funding the project. He reminded Mr. Bonner that these lakes had languished for lack of funding assistance. Mr. Sterling also stated that in a conversation with Mr. Steven Robinson, USAFE Project office of the COE, Mr. Robinson stated that USAFE Section 206 funding was a possibility. Section 206 funding provides a 65% federal allocation, with the remaining 35% being supplied by sponsor agencies such as the SJRWMD, FWC and local governments.

Mr. Sterling's letter stressed the point that the project needed to obtain authorization due to scheduled flooding of the planned Sawgrass Lake Water Management Area (SLWMA). The SLWMA Project is a part of the C-1 Rediversion Project which was already funded. Approximately 800 acres of the SLWMA could be used for depositing the spoil dredged from the lakes prior to start of construction for the C-1project. The COE Section 206 project could provide $5 million dollars to each lake for a total of $10 million dollars from the federal government.

On 24 November 1998 in a letter to this writer, Mr. Lothain Ager, with Aquatic Resources Enhancement Project of the FWC provided me with a copy of a "Conceptual Plan for Restoration of the Upper St. Johns River—Lakes Hell N' Blazes and Sawgrass and Protection of Lake Washington". The document had been recently released. Mr. David Cox, a marine biologist with the local FWC office in Melbourne was the author of the plan. With over 30 years experience on the Upper Basin of the St. Johns River, Dave was well qualified to write the plan. Scientific data from an earlier contract study performed by a team of scientists supported the restoration of these two lakes.

In a letter I wrote to Colonel Joe Miller, District Engineer for the U.S. Army Corps of Engineers (COE) in Jacksonville on 30 November 1998, I referred to Mr. Cox's conceptual plan. I expressed my frustration in the continuing delays in obtaining federal funding in order to proceed with restoring the first two lakes on the St. Johns. I pointed out to Colonel Miller that the St. Johns River had been awarded a federal designation as an "American Heritage River" at a ceremony in Jacksonville on 30 July 1998, less than 4 months earlier. I sought his support to move forward with the restoration project.

Less than two weeks later, on 4 December 1998, I wrote letters to U.S. Senators Connie Mack and Bob Graham and U.S. Representative Dave Weldon (local District 15 representative). I provided each legislator with background data relative to the SJRWMD, FWC, and SAVEs effort to secure needed federal funding. Likewise the SJRWMD had written a similar letter to these legislators. Each legislator responded favorably to both letters. Each one of them had contacted the COE, investigating the funding issues for restoring the subject lakes.

This writer worked with Mr. Steven Robinson, USAFE Project Engineer for the COE Section 206 funding issue, as did the SJRWMD. However, two years would pass before I received a letter from the COE Planning Division, dated 20 December 2000, stating the COE planned to solicit public input for *"a feasibility study to evaluate the effects of the Lakes Sawgrass and Hell N' Blazes restoration project"*. I responded by letter dated 29 December 2000. I reiterated SAVEs position had not changed since my initial request for this same study 12 years earlier (1988). I urged the COE to proceed without further delay. I stated I would provide a boat, if needed, for their agents to visit the lakes at any time.

On 18 September 2001, Mr. Hector Herrera, SJRWMD Operations department, provided this writer with a preliminary restoration schedule produced by the COE. The schedule included more detail of the milestone events. I noted that construction was planned to begin on 31 December 2002. After years of delay, I accepted this date with the feeling that at least I had something "in writing" although cost estimates had not been finalized at the time.

Restoration Cost Continue to Escalate

In March 2002, the FWC provided this writer with a COE cost-estimate to restore both Lakes Hell N' Blazes and Sawgrass. The total funding was $9,092,800. Again, using the USAFE 206 program, the federal government would provide 65% of the cost, while the SJRWMD, FWC, and local governments would provide 35% of the cost. It appeared things were about to get serious, but still more opportunities for tolerance would be necessary.

By October 2002, COE cost-estimates continued to escalate for the restoration project. Lake Hell N' Blazes restoration was now shown to cost $7,565,927. Lake Sawgrass restoration cost was shown to be $8,150, 400. The total cost to restore both lakes had now risen to $15,716,327. This was an increase of $6,623,527 dollars in less than 7 months. The following is a list of how these costs kept fluctuating (mostly increasing) over the years. Inflation I understand; these numbers go far beyond inflation. I can only conclude that the "left hand did not know what the right hand was doing". The numbers presented include restoration of both lakes:

July 1999	Prelim. Restoration Plan	$ 5.2 million
Nov 1999	Revised Prelim. Restoration Plan	$ 7.2 million
Jan 2002	Ecosystem Restoration Report	$ 9.9 million
Feb 2002	Revised Ecosystem Restoration Report	$10.3 million
Mar 2002	Revised Ecosystem Restoration Report	$ 9.1 million
Oct 2002	Revised Ecosystem Restoration Report	$15.7 million

In March 2003, Maurice Sterling, SJRWMD wrote yet another letter to Mr. Richard Bonner, Deputy District Engineer for the COE. Mr. Sterling stated his concern for the ever-increasing cost for restoring Lakes Hell N' Blazes and Sawgrass. He further stated that due to these significant increases in cost, he was not certain of the state's ability to undertake the project. A meeting was requested to discuss the financial issues.

Several alternatives were now under discussion; restore only one lake was an option. Which lake should get the restoration? SAVE did not support this approach; I stated if a decision was made to restore only one lake, then Lake Hell N' Blazes would be the best choice. The reason is obvious; to restore only Lake Sawgrass, minimal success would be provided. The polluted water from Lake Hell N' Blazes would enter Lake Sawgrass. SAVE continues to insist that both lakes must be restored to protect the potable water supply of Lake Washington. Both lakes would benefit by improvements in wildlife and fishery habitat.

Following the October 2002 revised cost-estimate increase to $15.7 million dollars, the COE dispatched the restoration report to the Mobile, Alabama district COE for review. The COE office in Jacksonville stated the Mobile district had performed more of this type of restoration. The COE indicated if the Mobile district could demonstrate savings in the project, the Jacksonville district would accept such findings. As a result of the Mobile review, costs were reduced from $15.7 million to $13.4 million.

Acting on behalf of SAVE, I initiated a search for local sponsors to assist the SJRWMD and FWC to locate the 35% match of $13.4 million dollars. I wrote letters, e-mails, and coordinated with local city mayors, the county commission, the COE, U.S. Senators Bill Nelson and Bob Graham, and U.S. Representative Dave Weldon. The responses were encouraging. The Florida Today newspaper responded with my story to the public.

The SJRWMD scheduled a meeting of all interested parties to be held on 18 August 2003 in Brevard County. The meeting was conducted by Steven Robinson, of the Jacksonville COE USAFE 206 Project office. In previous letters to the COE, the Brevard County Commission, Melbourne's City Manager, Mayor and Public Works Director had expressed their concern of potential impact to the city's public water utility.

The meeting was successfully concluded with a significant majority of city, county and state agencies committing to provide the 35% match to get the project underway. Steven Robinson closed his presentation by stating the sponsor's matching funds needed to be committed no later than 30 September 2003, only 6 weeks away. Matching funds were committed by the due date. Subsequently, sponsor entities were promised the restoration project would be underway in October 2004 (the start of the next COE fiscal year).

Restoration Saga—On Again, Off Again

On Thursday, 19 February 2004, at a meeting of the Upper Basin Recreation Advisory Committee, Hector Herrera, SJRWMD staff, announced that funding for all USAFE 206 COE projects had been stopped, including the Lakes Hell N' Blazes and Sawgrass project. This writer questioned the source and reason for the decision. Mr. Herrera did not know the specific reason; only that the SJRWMD received the notice from the COE. I was subsequently informed that the COE had a significant funding backlog of projects that had to be rescheduled.

The following day, 20 February 2004, I e-mailed all local partners who were committing financial funding for the project. I encouraged their committed funds be put in an escrow account, or set aside in a manner that would assure availability when we received a new date for start of the restoration project.

On 23 February 2004, I e-mailed Doris Marlin, COE office in Jacksonville. Ms. Doris replaced Steven Robinson. He was transferred to the Everglades project in south Florida. She was new on the job and knew little about Lakes Hell N' Blazes and Sawgrass restoration project and had not visited the project site. I scheduled an airboat tour of the project area; Ms. Marlin was accompanied by SAVEs Vice President Larry Gleason.

On 24 February 2004, this writer addressed the Brevard County Commission. I reported the COE office in Jacksonville had informed the SJRWMD that all USAFE 206 projects had been stopped due to a backlog in their own funding issues. The COE stated it would be two years (2006) before restoration work could begin on Lakes Hell N' Blazes and Sawgrass. Again, I recommended the board send a letter to the COE stating such a delay was going to impact the C-1 Rediversion Project.

This writer informed the board that if these lakes are not restored before the scheduled start of the C-1 project, either the C-1project would have to be delayed or SAVE would consider filing a federal lawsuit to delay start of the C-1project until the two lakes were restored. The problem: the area where waters from the C-1project would be diverted is the same area that would be used to deposit the spoil from the lakes being restored. Sufficient time would need to elapse, permitting the spoil site to solidify before the historic fresh water from the C-54 canal could be diverted into the St. Johns River.

On 11 March 2004, once again this writer contacted local, state, and federal officials requesting their investigation of the COE financial problems with the restoration project. Brevard County Commission chairperson Nancy Higgs wrote letters to the legislators previously identified. Commissioner Higgs clearly stated the board's objections to a two-year delay in the project. Responses were again received from the local, state and federal legislators, stating they were doing their best to get the restoration projects back on schedule.

On July 15, 2005, I received a fax from Jacksonville COE District Engineer David A. Tipple. The Jacksonville district of the COE had dispatched the restoration report; this time to the Charleston, S.C. COE office for review and comment. The Charleston COE response stated booster pumps were needed to move the "slurry" through an approximate two-mile pipeline (the distance from Lake Hell N' Blazes to the spoil site).

The Charleston COE district also questioned whether the 800-acre spoil site was sufficient in size. When their review was completed, the cost per cubic yard of spoil to be removed would increase from $3 per cubic yard to $6 per cubic yard. This figure would double the construction cost of the project. In addition, fuel costs would be significantly increased. The new cost estimate had risen from $13.4 to $21 million dollars.

Under the federal USAFE 206 program, the maximum funding available for restoring a lake is $5 million dollars. As stated earlier, since two lakes require restoration, the USAFE 206 program could provide $10 million dollars of the funding. This would now leave $11 million dollars for the SJRWMD, FWC and local government partners to pay. Recently, I was informed that the Jacksonville COE district has removed these lakes from consideration under the USAFE 206 program due to the increased cost; therefore that program was no longer an option.

On 4 April 2006, Florida Today Environmental Writer Jim Waymer's headline story read: *"Officials Kill Plan to Dredge—St. Johns Lake Project too Costly"*. Mr. Waymer reported the comments of Hector Herrera, Senior Project Manager with the SJRWMD: *"As far as the district is concerned, we are no longer pursuing the project. If another agency wants to pick up the project and do it on its own, that's their choice"*. Mr. Herrera further commented: *"The Army Corps of Engineers was able to contribute a maximum of $5 million per lake, but the water management district would have had to sacrifice other projects to cover the rest. That project would have taken money from other high priority projects"*.

Assistant City Manager of Melbourne, Mr. Howard Ralls responded: *"The water plant on Lake Washington's western shore supplies customers in Indialantic, Melbourne Beach, Satellite Beach, Indian Harbor Beach, Palm Shores, Melbourne Village, and some unincorporated areas"*. Perhaps a misprint, but the city of Melbourne was not listed in the article. Melbourne is the largest consumer of water treated via the treatment plant.

Mr. Ralls also stated: *"Those (dredging) areas* (referring to Lakes Hell N' Blazes and Sawgrass) *are upstream of our water plant. They could potentially impact water supply. I'm not aware of a good technical evaluation of that project as it relates to water supplies"*. Mr. Ralls stated that City Manager Jack Schluckebier wants to investigate the situation further, perhaps by commissioning an independent study. Melbourne City Councilman Richard Contreas stated: *"the situation is a ticking time bomb; the water supply could be just on the edge of being mucked"*.

On 10 April 2006, this writer's "Letter to the Editor" of Florida Today was published regarding the issue of restoring Lakes Hell N' Blazes and Sawgrass. My letter entitled: *"Dredging Project Must Get Financial Support"* is quoted in part: *"I will work with our partnership, including the water management district, to restore these lake fisheries and protect the potable water supply of Lake Washington. This is a "must do project" that keeps falling by the wayside. A plan will be formulated seeking financial support through local and state officials. The FWC should be the lead agency. This agency has the expertise to get this job done. Annually, over $20 million dollars are spent on sand to restore our beaches. Surely, a one-time cost of $10 million dollars can be allocated to restore the first two lakes on the St. Johns River"*.

Wanted: Financial Support and a Lead Agency

Space is limited to 200 words in a Letter to the Editor of Florida Today. This writer is less restricted in the writing of this book. I feel compelled to state that the funding needed is a joint responsibility of the FWC and the SJRWMD. Municipalities receiving potable water from Lake

Washington must be willing to contribute proportionally to the funding needs of the project. Is the restoration of Lakes Hell N' Blazes and Sawgrass more important to restore a fishery or to protect the health of citizens who use the potable water from Lake Washington? A reasonable person would agree that protecting the health of citizens far outweigh the health of a restored fishery.

One may conclude that the SJRWMD should be the lead agency. Since both issues, that of restoring a fishery, or that of protecting the potable water supply of the lake is significantly important; my recommendation would be to pursue the restoration as a joint effort, supported by Brevard County Government and affected municipalities. If nothing is done, the citizens could lose both benefits; the fishery will remain dormant and the citizens could pay in the end via poor health and/or higher cost for treatment of the water supply. The partnership must expand to include all communities receiving potable water from Lake Washington. In addition, each partner member should work with local and state legislators to obtain additional funding.

SAVE Calls an Urgent Meeting

On May 25, 2006 this writer organized and opened a meeting with the key agency personnel from the SJRWMD, FWC, county government, the city of Melbourne, and several aides to local and state elected officials. The open-ended meeting provided one significant accomplishment. The SJRWMD would investigate the possibility of an updated study to determine the extent of sediment being transported into Lake Washington. Hopefully, the amount of sediment that has already reached the lake will be included in the study.

Maurice Sterling, SJRWMD stated that the original spoil site planned for use in restoring the lakes is now an unlikely location as the site has recovered to a functioning wetland treatment area. Another issue yet to be resolved if the restoration is to happen, is who will take responsibility as lead agency for the project. That issue may be resolved by the results of an anticipated study.

Brevard County Commissioner Sue Carlson, who serves a Chair of the St. Johns River Alliance, dispatched a letter to the SJRWMD on 2 June 2006. Commissioner Carlson stated in the letter that Maurice Sterling of the SJRWMD technical staff would need to conduct a more detailed sediment transport analysis to determine the actual rate of sediment deposition from the upstream lakes into Lake Washington.

The SJRWMD responded to Commissioner Carlson's letter with a commitment to perform the needed study in the next fiscal year (which starts October 2006). No further action is anticipated until the findings of the study are discussed, hopefully in the early fall of 2006.

As I conclude this chapter, as President of SAVE and a board member of the St. Johns River Alliance, I remain committed to the restoration of Lakes Hell N' Blazes and Sawgrass. I have contacted the SJRWMD staff and discussed the need to obtain an easement with sufficient storage capacity to contain the sediments. Also, the FWC performs restoration projects around the state. This agency, supported by the SJRWMD and the partnership, could restore these lakes at much less cost if an easement could be obtained near the lakes.

If necessary, acquisition of viable disposal site acreage is available through immanent domain proceedings. This process should be considered should the private landowner nearest to the lakes be unwilling to sell an easement to the SJRWMD. Immanent domain proceedings should only be undertaken as a last option to secure an appropriate site for deposit of sediments. However, protecting the drinking water supply for 150,000 residents would render such action to be in the public interest.

More Fishing Trip Memories—In the summer months of any year, rainfall from storm events present great fishing opportunities. On the St. Johns River, especially in the Upper Basin, several inches of rain within a 24 hour period can provide excellent bass fishing near areas where runoff enters the river. Such events happened almost every summer. I especially recall the summer of 1980. At dawn on a Saturday morning, I launched my boat from SR 520 at the river. I traveled to an area approximately 10 miles north of SR 520. A "natural" low-lying area was offloading a significant amount of water into the river. This was going to be a great day and no other boat was in the area. Within the first hour of fishing, I caught and released approximately 25 bass. Bass weighing 2 to 4 pounds were "schooling" near the runoff. For two hours, the feeding frenzy continued. I discovered a problem. My artificial worm supply was running out; I improvised by re-hooking "damaged" worms. This process usually works when schooling bass are present. However, fishing began to taper off. I theorized it was because the half-the-normal length damaged baits were not appealing to the bass.

I cranked my motor and departed for a trip back to a tackle shop. I needed some new plastic worms. One hour later I was back "on my spot". I fished the next 2 hours and never got a "bite". Either the bass had finished their morning meal, or they had relocated. I cranked my motor and headed home. I had enjoyed a great and successful day on the water. I have a bumper sticker on my boat trailer which reads: *"I SAY WE WORK TWO*

DAYS AND FISH FIVE DAYS". Another one I have seen reads: *"A BAD DAY FISHING IS BETTER THAN A GOOD DAY AT WORK"*. All anglers would agree with both statements.

CHAPTER V—

PROPOSED SABAL HAMMOCKS PROJECT

In this chapter, the writer will describe a developer's dream project. Most assuredly, a highly profitable investment, with amenities abounding, would have appealed to hundreds of citizens. If one's favorite hobby is golf, why not live in a residential golf-course community. The historic St. Johns River would be "your back yard". Nearby, the revitalized city of Cocoa provides a variety of specialty shops, excellent eateries, and tree-lined "downtown" streets. Five miles east of the community, Interstate I-95 is accessible. Cocoa Beach and the Atlantic Ocean are only a 30-minute drive. To the west, Florida's most famous vacation attraction, Disney World is only an hour's drive.

What happened to the landowner's dream project? Simply stated: The Sabal Hammocks residential/golf course community was planned for development in the wrong location. The multi-million dollar project was to be located behind a non-permitted 18,000-foot dike in the historic floodplain of the St. Johns River (Lake Poinsett) in Brevard County. In 1989, SAVE became very concerned with potentially negative environmental impacts of the proposed development.

Excluding the volumes of legal documents generated since 1990, my personal files on this project once filled a two-drawer filing cabinet. The reader will discover how SAVE won the support of thousands of Florida citizens, including the Governor and Cabinet to stop further degradation of the water quality in the St. Johns River. The mission of SAVE remains one of protecting and restoring an environmentally endangered St. Johns River.

Background/Ownership—Sabal Hammocks Property

In 1845, Florida became a state. By virtue of its sovereignty, Florida

3-mile dike at proposed Sabal Hammocks Project—view shows state's claim of sovereign lands at the site

becomes owner for the benefit of its inhabitants of all lands under bodies of navigable water within its territorial limits.

In 1850, federal government surveyors meandered the St. Johns River; however, Lake Poinsett was not a part of the meandering survey. Meander lines serve only to approximate the general "curves or winding"of the shoreline and were not intended to locate the ordinary high water line (OHWL) boundary. All lands below the OHWL are held in trust for the public.

In 1856, The State of Florida received a patent of "swamp and overflowed" lands from the U.S., pursuant to the Swamp and Overflowed Lands Act of 1850. Established case law holds that federal swamp and overflowed land patents do not convey or affect the public status of navigable lakes and rivers. Levies and dikes are typically used to drain swamp lands that are located **above** sovereignty lands. The capability to construct dikes and levee systems on sovereignty lands does not convert lands to a "swamp and overflowed" status.

In 1906, the State of Florida sold large tracts of swamp and overflowed lands to the current owner's predecessor in title, the Florida Coastline Canal and Transportation Company. The Florida Supreme Court law states that recipients of state swamp and overflowed land deeds were on notice that navigable waters were not included, so that those recipients lack "any moral or legal claim" to navigable lakes or rivers not explicitly excluded from deeds.

In 1934, F. Burton Smith bought the property addressed in this chapter, from the Florida Coastline Canal and Transportation Company, with notice that they had "no moral or legal claim" to lands under navigable water bodies.

In 1955, Chapter 30542—Laws of Florida was enacted. Chapter 30542 concerned only permits from the flood control district. It did not bear on the question of sovereignty land determination for the Smith property. It did not seek to establish the OHWL delineating sovereignty lands from privately owned uplands. Chapter 30542 "became void" on 1 June 1959 and was not law at the time of the construction of an illegal dike by Smith in 1973.

During the period of 1960 to 1972, tracts of land east of the Smith property were filled, including the residential canal homes known as Lake Poinsett Shores. The filled residential subdivision lands are not adjacent to Smith's property and are not waterward of the Smith's dikes. These lands were filled over 35 years ago. Lake Poinsett Shores homeowners are not the parties who violated state statutes; their lands are not at issue. In Smith's case, he is the party who violated state statues.

Mr. Smith was ordered to remove the illegal and obstructing dike from state lands which would restore the floodplain impounded by the dike. The landowner, F. Burton Smith, refused to comply with the order. The reader should understand the state of Florida does not pay for lands underlying navigable waters.

Illegal Dike Construction—Smith Property at Lake Poinsett

In 1973, F. Burton Smith constructed the illegal dike at 12.1 foot el. According to Smith's own surveyor's finding in 1977, the OHWL was established at 13.0 foot el. Until the 1980s, few "public eyes" observed the abuse of sovereign submerged lands; still fewer reported the abuse to the proper agency. In the absence of a "permit application", the Department of Environmental Regulation, in most cases, was unaware of such gross violations of law. However, during the 1973 construction of the dike at Smith's property, Mr. Frank Demsky, a Marine Patrol Officer indicated to Smith that the dike was landward of the OHWL and did not need a permit. Mr. Demsky was not qualified to render such a decision.

Landowner David Smith would later state in official pleadings: *"Demsky...saw the 1969 dike (constructed waterward of the 1973 dike) when he investigated the construction of the 1973 dike, but he thought it was a natural berm"*. It was because Demsky mistook the 1969 dike for a natural berm that he concluded the new 1973 dike was landward of the OHWL. Smith never informed Demsky of his mistake.

In 1976, the Department of Natural Resources notified the applicant (Smith) in writing that: *"field investigation has shown the dike location would have required permits from the state at the time of original construction...the dikes exist as violations of state statues ; the files are being forwarded to our enforcement section"*. Subsequently, the Department of Environmental Regulation ordered Mr. Smith to breech the illegal dike in 21 locations to allow the lake water to refill the interior portion of the property.

In 1981, a field report filed by Mr. Ed Carr of the SJRWMD enforcement division stated: *"this area can no longer be considered a closed system, because the low cuts that were placed in the outside dike will allow a direct connection to the St. Johns River...during high water"*. A bulldozer operator informed Mr. Carr that he was instructed to cut these low places in the dikes at water level for water movement in the marsh areas. Mr. Carr's field report further stated: *"Mr. John Dyal, Project Manager for the property verified that information"*.

In 1982, the Paradise Fruit Company applied the Marketable Records Title Act (MRTA) to quiet title of a large portion of Lake Poinsett "bottom lands". However, MRTA could not be used to convert sovereignty lands into private ownership. Subsequently, Paradise Fruit Company's claim of ownership was overturned and MRTA was subsequently rescinded by the state of Florida.

In 1986, David Smith, son of deceased F. Burton Smith argued with the Department of Environmental Regulation and the SJRWMD that the 1973 dike was not illegal. Mr. Smith apparently relied on a 1955 flood control project act rather than the explicite information provided to Smith in 1976 by the Department of Environmental Regulation, the Florida Game & Fresh Water Fish Commission, the SJRWMD and the U.S. Corps of Engineers.

From 1976 to 1982, these agencies told Smith repeatedly that his dike was impounding floodplains, was below the OHWL and was illegal. Mr. Smith stated that the dike qualified for a maintenance exemption based on its being an agricultural operation. Potentially, such an exemption would have allowed Smith to legally fill in the "ordered breeches" of the 1973 dike. No written record of any agreement to allow filling in the breeches of the dike was ever located. Subsequently, Smith would argue that he received "verbal" authorization from the SJRWMD. The Department of Environmental Regulation would not accept Mr. Smith's argument.

Planned Purchase of Smith Property

In 1987, the SJRWMD was engaged in a partnership with Brevard County to purchase Mr. Smith's property. On 19 March 1987, Mr. John Hankinson, Director of Land Acquisition for the SJRWMD, wrote a letter to Mr. Jim Swann, a member of the SJRWMD Governing Board. Mr. Swann had requested the history of the SJRWMD involvement with Brevard County relative to the partnership. The following paragraphs address the history, as stated in Mr. Hankinson's letter to Mr. Swann:

"As originally envisioned, the acquisition sought to acquire roughly 1,900 acres...of Smith's property. I visited the property at the request of Mr. Tom Lawton and Ms. Teresa Kramer who then worked for Brevard County....Also present was Gwen Heller, a land acquisition agent for the County. The property was under appraisal at the time...Gwen Heller advised that they had some "problems" with the appraisals, that they were too low...this comment understandably made me very nervous...

Our next contact from Brevard County about this parcel was the receipt of a Resolution from the County Commission asking the District's participation in the project. The Board (SJRWMD)...directed staff to begin "meaningful negotiations" with Brevard County with regard to sharing the cost of the wetlands. I sent a letter dated October 22, 1986 to Gwen Heller...inviting the start of these negotiations. I never received a response from my letter. When I did speak with her, I was told that...they were in the process of securing a third appraisal...I was given the impression that this project may have problems and that it was unnecessary to discuss a joint acquisition at the time.

A week to ten days prior to the District's March Governing Board meeting, I was discussing some other property with Ms. Denise DeVito with Brevard County...and inquired as to the status of the Lake Poinsett property. She told me that an agreement had been reached on that property, and that it was going to the County Commission on...March 10, 1987, the day before our Governing Board meeting. I expressed surprise, stating that I thought this was to be a joint acquisition, and that I had never been contacted by Brevard County to work out a cooperative arrangement.

Ms. DeVito stated that it was her understanding that (after the County had entered into the agreement) they would be coming to ask the District for its share of funding for the wetlands. I advised her...we have certain...requirements governing our land acquisition process.....I also told her that I did not feel the Board would be favorably inclined to committing funds to a project where their agent had not been involved in the negotiation.

On March 13, 1987, I met with Teresa Kramer and Denise DeVito....I had not been aware that Ms. Kramer had become the landowner's agent in this transaction...they presented the deal that was entered into by the County....I was surprised to discover that 750 acres of the upland property had been excluded from the project...We had no role in appraiser selection or quality control....It appears from a cursory review that at least as to some of the property a higher than market value was paid...As I understand it, "pasture land" outside the dike was valued at $1,375 an acre. (Writer's note: some 900 acres were in the open waters of Lake Poinsett—submerged sovereign lands held in public trust)...

The landowner was allowed to retain the perimeter dikes, which would negate the ability to breach these dikes and restore the area...It is my understanding that the property retained by the landowner will be subject to re-zoning and be considered for further development, possibly a retirement/golf course development....

My recommendation which I shared with Denise DeVito, Teresa Kramer, and Sue Schmitt (county commissioner)...we may be willing to participate...in restoration activities and other water management concerns....I feel the County, and perhaps Commissioner Schmitt individually stands to suffer embarrassment if this arrangement is put under the microscope in a public debate...It well may be that the County has obtained what it needs in an arrangement that is quite suitable to their purposes. It does not however, appear to meet the requirement the SJRWMD Board would impose on a land acquisition project".

In a letter dated 10 June 1988, to Mr. Rick Enos, Chief Senior Planner for the County, Teresa Kramer, of Environmental Site Design, forwarded a *"package of information which constitutes our application for preliminary development plan approval for the proposed Sabal Hammocks at Lake Poinsett...This project is located on 716 acres which is currently zoned AU"*.

In another letter dated 18 July 1988, to Mr. Rick Enos, from Teresa Kramer, she provided *"our response to the County Staff comments on the Sabal Hammocks at Lake Poinsett P.U.D. submittal"*. (Writer's comment: P.U.D. is an acronym, meaning Planned Unit Development)

SAVE St. Johns River Group Investigates Proposed Development

In 1989, Mr. Smith applied for a new Management and Storage of Surface Water (MSSW) permit from the SJRWMD for the proposed Sabal Hammocks project. SAVE informed the SJRWMD of our objections to issuance of the MSSW permit and the project as proposed. However, the district planned to award the permit as no rule existed for denial. (Note: Pumps had been utilized for over 30 years on this illegal dike.)

In a letter to Commissioner Sue Schmitt, dated 29 January 1990, this writer expressed SAVEs concerns about the proposed Sabal Hammock project. I informed the commissioner that SAVE had been provided a tour of the property being considered for a subdivision. I also stated that recently I was informed of plans for a golf course on the property.

Subsequently, I received a telephone call from David Smith, the owner/developer. Mr. Smith stated that the water supply from the "man-made lakes" on the property was not suitable for reuse and would be offloaded via a pump station, into the St. Johns River. My letter concluded with the following statement: *"SAVE could not support any encroachment that will render our already fragile environment of Lake Poinsett…with any additional damage to the water quality…such a project must be revisited and some alternative use made of the property".*

In 1990, Mr. Smith formally requested a jurisdictional boundary from the Department of Environmental Regulation. The Department contended that Smith's filling in of the previously ordered breeches of the dike was not exempt because the property was not being managed for grazing cattle, but was instead being proposed for development as a subdivision and golf course. Mr. Smith withdrew his request for the jurisdictional boundary.

On 6 July 1990, Assistant Brevard County Attorney Kenneth C. Crooks, wrote a letter to Nancy Barnard, Office of General Council for the SJRWMD. Subject of the letter was: Sabal Hammocks—Links at Lake Poinsett Development, David Smith Property. In Mr. Crooks letter, he stated: *"please be advised that it is the current intention of the Board of County Commissioners of Brevard County, Florida, to accept responsibility for the operation and maintenance of the stormwater utility system associated with the above listed Planned Unit Development and associated public golf course project, pursuant to the terms and conditions of the permit issued by the St. Johns River Water Management District".*

This writer verified that the subject had never been presented to the Board of County Commissioners for discussion. The SJRWMD was informed that Brevard County had made no such commitment for operation and maintenance of the stormwater utility system.

SAVE Makes It Official—Files Legal Challenge against Developer and the SJRWMD

On 16 July 1990, the SJRWMD notified their Governing Board that SAVE had filed a petition challenging the district's issuance of the MSSW permit for the proposed Sabal Hammocks project. SAVEs petition was forwarded to the Division of Administrative Hearings. An Administrative Hearing was scheduled for February 1991. The road to Florida's Governor and Cabinet is set in motion.

On 17 July 1990, in a letter to Brevard County Commissioner Schmitt, this writer advised the commissioner that SAVE had filed a legal challenge to deny issuance of the MSSW permit, in support of the golf-course/residential project.

I reminded Commissioner Schmitt that she had recently promised this writer, that if SAVE challenged the project, she would pull the County out of the golf-course agreement. I also reminded the commissioner that our support base was expanding and included numerous state environmental groups and a major national corporation (Bassmasters, Inc.) in opposing the project. Subsequently, in SAVEs appearance before the Governor and Cabinet, arguments were presented that sovereign (state owned) lands were included in the proposed development.

In a Florida Today article published on 7 August 1990, Commissioner Schmitt stated she would, in an upcoming Brevard County Commission meeting, request the county cancel any agreements relative to the golf-course at Sabal Hammocks. I quote from her statement in the article: *"Brevard County should back out of a proposed west Cocoa golf course because of protests from environmentalists and sportsmen"*. In the same article, I stated: *"the property contains old wetlands that should be restored to their original state. The project appears to violate the SJRWMDs history of steering development away from environmentally sensitive areas"*.

On 7 September 1990, Commissioner Schmitt requested and received approval from the Brevard County Commission to remove Brevard County as a participant in the development and operation of the golf course.

The scheduled Administrative Hearing in Brevard County between the plaintiff SAVE vs. defendants SJRWMD and developer David A. Smith was held in February 1991. SAVE did not prevail at the hearing, primarily due to the poor performance by our legal council. Mr. Dennis Auth, SAVE consultant, had provided significant data to SAVEs attorney prior to the hearing. Nothing of consequence was presented to the hearing officer for consideration on SAVEs position, while numerous documents were placed into the record by the SJRWMD and council for the developer. It was not a good day for SAVE. The search for a replacement pro-bona attorney was initiated on that day.

On 26 May 1991, Florida Today Outdoor Writer Bill Sargent, published an "unusual" article on an upcoming "boat parade" sponsored by SAVE. On 1 June, 47 boats pulled by trucks and/or vans participated in the prearranged, police-escorted, parade route through Central Brevard County. The parade route began at the St. Johns River off SR 520, proceeded east onto U.S. Highway 1, north onto SR 524, and west onto SR 520, ending at the river two hours later.

SAVE supporter Wanda Tester awaits start of boat parade to protest Sabal Hammocks Project

Boats were equipped with large signs with various statements of opposition to the proposed Sabal Hammocks project. The route drew considerable public attention. Many citizens along the route cheered us on; the sound of car horns, accompanied with "thumbs up" acknowledged our cause; it was a very successful "eye-catching" event. The next day, 2 June, SAVE held its Sixth Annual Benefit Day, raising over $6,500 in support of our legal expenses.

SAVE Support Base Mobilizes for Potential Appearance before the Governor and Cabinet

During June 1991, SAVE worked very closely with our team members: Environmental Consultant Dennis Auth, of Jacksonville; Water Quality expert, Dr. Forrest Dierberg of Melbourne; and Florida Fresh Water Fish & Game Commission Marine Biologist David Cox. I also contacted Attorney Tom Tomasello, of Tallahassee, who agreed to review our case and contact this writer as soon as possible.

This writer continued to work with state conservation and environmental groups in our effort to stop the proposed Sabal Hammocks Project. Eleven statewide environmental/conservation groups committed to support SAVE in preparation of our appearance before the Governor and Cabinet. SAVE membership was at its highest level with 30 organizations from three counties (Brevard. Orange and Indian River). SAVEs support base totaled over 3,000 citizens.

SAVE continued to receive considerable "media coverage" in the Orlando Sentinel and Florida Today. Other newspapers including the Tampa Tribune, Daytona Beach Journal, Tallahassee Democrat, and the St. Petersburg Times published numerous articles on this unfolding story.

Legal Issues Begin in Earnest

On 2 July 1991, the Hearing Officer filed her Recommended Order in SAVEs case against David Smith, and the SJRWMD. SAVE failed in this initial attempt to deny the MSSW permit previously approved by the SJRWMD. As this writer, our consultant and expert witness were unable to contact our attorney, SAVE consultant Dennis Auth prepared "Exceptions to the Order" and submitted same to the SJRWMD Governing Board Chair, Ms. Sandra Gray. SAVEs arguments were not accepted; reason stated: *"Mr. Auth is not council for SAVE".* Following the final hearing by the SJRWMD Governing Board and approval of Smith's MSSW permit, SAVE terminated a "broken" agreement with the original attorney.

On 2 August 1991, on behalf of SAVE, without the benefit of an attorney, I forwarded a 46-page package to Governor Chiles for review by his staff. The basis of the two-page cover letter was to inform the Governor of SAVEs position relative to the proposed project, in anticipation of appearing before the Governor and Cabinet in the near future. The enclosures attached to my letter represented SAVE consultant Dennis Auth's research documents, news articles, list of major support groups around the state, and a map of the proposed project's location behind an illegal dike.

On 1 October 1991, Attorney Tom Tomasello with the law firm of Oertel, Hoffman, Fernandez & Cole, P.A., located in Tallahassee, notified this writer that the firm would represent SAVE on a pro-bona basis throughout any future legal proceedings. Things were about to get more exciting.

The Florida Game and Fresh Water fish Commission dispatched a letter to Ms. Carol Browner, Secretary, Department of Environmental Regulation, dated 3 October 1991. Colonel Robert Brantly, Executive Director for the GFC, informed the Secretary of the commission's three major concerns about the proposed Sabal Hammocks project. These concerns were:

"1) transforming marshlands into residential development; 2) the correctness of using agriculture as the pre-development condition for purposes of determining allowable discharges into the St. Johns River, and 3) whether the project is contrary to the State Water Policy and the District's marsh restoration program, whereby residential development may be allowed in the floodplain of the St. Johns River". (Writer note: the Florida Game and Fresh Water Fish Commission was now providing two attorneys in support of SAVEs case.)

On 4 November 1991, this writer forwarded a letter to Department of Natural Resources Secretary Carol Browner. I solicited the Secretary's support for SAVEs upcoming appearance before the Governor and Cabinet. I identified 10 major state environmental groups, plus the State League of Women Voters, who fully supported SAVEs effort to overrule the SJRWMD MSSW permit awarded to David Smith, and protect the marsh/floodplain and sovereignty lands associated with the proposed project.

On 21 November 1991, SAVE consultant Dennis Auth wrote a letter to Mr. David Maloney,

General Council, Office of the Governor. Mr. Auth provided an 8-page synopsis of SAVEs case. Mr. Auth had attended a cabinet aides meeting the previous day where Mr. Maloney had raised a number of points on the subject case. In Auth's letter, he provided explicit details dating back to the 1978 enforcement action by the Department of Environmental Regulation to restore the lower dike constructed in 1973 without required permits.

SAVE St. Johns River, Inc. Case Heard by Governor and Cabinet

On 26 November 1991, SAVE appeared before Governor Chiles and the Cabinet. Following a lengthy hearing regarding the SJRWMD approved MSSW permit, a 3-3 tie vote occurred (Cabinet member Betty Castor was absent). Two weeks later, in a second appearance before Governor Chiles and the Cabinet, Betty Castor voted in opposition to the MSSW permit; however, Comptroller Gerald Lewis changed his vote to favor the permit. The vote was 4-3 to leave the permit, as approved by the SJRWMD in place.

During arguments by SAVE attorney Tom Tomasello, regarding the presence of sovereignty lands on the project site, Attorney General Bob Butterworth requested an investigation. After discussion among the Governor and other members of the Cabinet, an order was issued for the Division of State Lands to visit the Sabal Hammocks project site and make a preliminary assessment of the OHWL and report the findings to the Governor and Cabinet within two weeks.

By letter dated 20 December 1991 from the writer, SAVE authorized Attorney Tomasello to appeal the Governor and Cabinet 4-3 vote, sustaining the MSSW permit, to the First District Court of Appeals. Subsequently, the court ruled 2-1 to affirm the Governor and Cabinet vote.

Also, on 20 December 1991, Percy Mallison, Director of Division of State Lands, notified Mr. Bram Canter, attorney for David Smith that *"based on a recent site visit by members of our Bureau of Survey and Mapping, as well as water gauge data from the U.S. Geological Survey, we believe the ordinary high water line is not below the 13 foot contour. Consequently, we recommend that your client not undertake any activities below that line until we can resolve this matter to everyone's satisfaction"*.

SAVE Brings About Statewide Floodplain Protection Policy

As a result of SAVEs appearance before the Governor and Cabinet, another significant event was addressed. The issue: consistency between all five water management districts in their policy of addressing floodplain protection on all rivers in Florida. The Governor and Cabinet ordered the Department of Environmental Regulation to create a statewide floodplain protection policy that would incorporate consistency among all five water management districts. The order further stated that a preliminary status report be provided to the Governor and Cabinet at the 8 May 1992 cabinet meeting.

On 8 May 1992, the Department of Environmental Regulation provided the Governor and Cabinet the status of the statewide floodplain protection policy. I quote, in part, from the status report: *"...considering the matter of SAVE the St. Johns River, Petitioners v. St. Johns River Water Management District (SJRWMD) and David A. Smith, Respondents...the commission...agreed that protecting the floodplains in the State of Florida should be the highest of the highest priority to the water management districts...*

Florida has over 12,000 miles of streams and rivers, more than 7,700 lakes, and 8,500 miles of coastline....Several initiatives are underway to develop the State Water Management Plan, with target date of October 1992 for an initial draft and November 1, 1994 for its full development". Indeed, SAVEs legal challenge brought about the development of consistent statewide floodplain protection policy.

Governor and Cabinet Issue Final Order for OHWL Survey

In 1992, the Governor and Cabinet issued a final order directing the Division of State Lands to investigate the state's interest in the matter of David Smith and the proposed Sabal Hammocks project. State Lands surveyors Terry Wilkinson and Rod Maddox, with technical support from the SJRWMD would eventually perform the OHWL survey in 1994 to determine the extent of sovereign lands at the project site.

SAVE Obtains Major Corporation Support

At the request of this writer, Bassmasters Inc., with corporate headquarters in Montgomery, Alabama, supported SAVEs position of protecting the floodplains of the St. Johns River. The corporation publishes its national magazine "Bassmasters" every other month. Membership in Bassmasters, Inc. is over 650,000 (38,000 in Florida). Dr. Al Mills, Natural Resources Director for Bassmasters Inc. dispatched a two-page letter to the Governor and Cabinet recommending denial of the proposed Sabal Hammocks project.

Mr. Robert Montgomery, Senior Writer for Bassmasters magazine traveled to Brevard County, meeting with this writer for a tour of the problem areas of the St. Johns River. In the December 1992 issue of the magazine, a three-page story was printed regarding Mr. Montgomery's observations on the river. His well written presentation left no doubt how Bassmasters Inc. viewed the negative impacts of massive pumps offloading poor water quality into the St. Johns River.

Brevard County Government Orders Developer to Return Advanced Funding

On 21 February 1993, Florida Today published an article regarding $276,000 Mr. Smith had been advanced for planning and design of the golf-course at the proposed Sabal Hammocks project site. SAVE and other citizens has pressed the Brevard County Commission to reclaim the advanced funds. They were included in a bond program established for the golf-course construction. Since the county had "backed out" of the agreement and withdrew its support of the golf-course, Mr. Smith would receive no further funding.

SAVE believed Mr. Smith should return any funds drawn from the bond account. In the article, Commissioner Truman Scarborough stated: *"We have a legal obligation to enforce this, but hopefully he will go ahead and pay it"*. Mr. Smith had previously received additional time, and again requested the commission to hold off until the legal issues were resolved. The article concluded with a statement from this writer: *"I think the county has been cooperative by giving (Smith) extra time already"*. The county placed a lien on property owned by Mr. Smith until the issue was settled.

SJRWMD Seeks Clarification of Brevard County re: Maintenance Entity

On 31 March 1993, Mr. Jeff Elledge Director of Resource Management, SJRWMD dispatched a letter to Commissioner Karen Adreas, Chair of the Brevard County Commission. Mr. Elledge stated he had relied upon letters from Mr. Peter Wahl dated 20 June 1990, and Ken Crooks dated 6 July 1990, to confirm Brevard County's intent to be the entity responsible for maintenance requirements stated in the permit granted to Smith on 9 July 1990. The commission was advised that if Brevard County decides not to participate, an alternative maintenance entity would require the permit be modified accordingly. The issue is mute since Mr. Smith has not attempted to obtain required county permits for any development of the property.

State Survey Revealed Brevard County Purchased State Lands

During the state OHWL survey performed on the proposed project in 1994, an interesting result was easily documented by the state survey regarding Brevard County's purchase from Mr. Smith in 1987. The county had paid Mr. Smith $1.3 million for 1,100 acres. The 1994 OHWL survey established that approximately 190 of the 1,100 acres had "clear" title. At the time of the purchase, the Brevard County Commission was informed by the chief county attorney that the county would not be able to purchase "title insurance" for acreage located outside the diked property, in the open waters of Lake Poinsett.

The Brevard County Commission utilized Beach and Riverfront funds for the purchase of 1,100 acres of David Smith's property. This writer located the public records of the "county purchase" and delivered them to a county commissioner. I requested the matter be addressed by the Board of County Commissioners.

This writer recommended the contracts be cancelled for one simple reason; Mr. Smith did not own the submerged lands under the navigable waters of Lake Poinsett on the St. Johns River. I recommended the county recover approximately $760,000 of taxpayer money for the 910 acres of submerged state lands. I never received a response; the issue was never presented to the Brevard County Commission. I was not surprised; this would have been an embarrassment to the Commission and could have taken years to settle. In an earlier part of this chapter, Mr. John Hankinson, former Director of Lands Acquisition for the SJRWMD expressed his surprise that Brevard County government avoided the SJRWMD regarding the purchase of 1,100 acres of Mr. Smith's property.

State Establishes Ordinary High Water Line at Proposed Sabal Hammocks Project

In 1994, the Division of State Lands conducted an OHWL survey at Mr. Smith's proposed Sabal Hammocks project site. The results of the survey concluded that the OHWL to be a line approximating the 13.8 ft. el. contour line. Subsequently, the state agreed to accept Mr. Smith's own 1977 OHWL survey which placed the OHWL at 13.0 ft. el., which resulted in the state's claim of 270 acres of submerged lands inside the illegal dike. The investigation process would continue, as would discussions with Mr. Smith relative to a redesign or other alternative use for the property. Regarding the state's claim of sovereign lands inside the 1973 dike, by law, the issue would be decided in court at a future date.

In August 1995, following the replacement of four of the seven state Cabinet members, Mr. Smith's team attempted to persuade the "new" Cabinet members to allow a disclaimer of sovereign lands on the proposed Sabal Hammocks project. SAVE also was in contact with these new Cabinet members, assuring each of them that the findings by the Division of State Lands should not be reconsidered. We also advised the new Cabinet members that Attorney General Bob Butterworth wanted the issue to go forward to a quiet title claim of the affected sovereign lands.

SAVE Organizes Tour of Dikes by State Officials

On 13 October 1995, SAVE provided the state Cabinet aides an "air-boat tour" of the proposed development site and the illegal dike planned for use to keep the development dry. On 14 October, the Florida Today printed a significant article on the Cabinet aides visit to the proposed project site. A front page photo of the tour boat, fully loaded with officials, was shown enroute to the 1973 dike.

The tour Captain searched for a clear stop-off point along the 18,000 foot dike to view the inland marshlands from atop the dike. Another photo in the article captured the Cabinet aides atop the dike where this writer presented a large map of the project, which clearly displayed over 150 homes and part of a golf-course to be constructed on the state's claim of 270 acres of sovereign (public) lands. In the article, I stated: *"All I can do is encourage you to be honest about what you see out here when you report back to your leaders"*.

SAVEs attorney was present for the tour, as was Brevard County Commissioner Nancy Higgs, marine biologist Dave Cox, with the Florida Freshwater Fish & Game Commission. Also present was a Florida Today reporter and a photographer. General comments by some of the cabinet aides supported SAVEs position to deny development on these historic marshlands. One cabinet aide stated: *"If the dike was not here, there would be 5 feet of water where we are standing"*; another cabinet aide responded: *"Why would anyone want to live "down there"?* (referring to the watery marsh). Approximately 15 officials participated in the tour.

Following the tour, as SAVE members discussed the issues with the officials; Mr. Smith appeared on the scene, uninvited by SAVE, who sponsored the tour. I considered his uninvited presence to be in poor taste, however, I did not ask him to leave.

In late November and early December 1995, the Governor and Cabinet received letters of strong support to proceed to quiet title of any sovereign lands included in the proposed Sabal Hammocks project. The Brevard County Commission, the Florida Freshwater Fish & Game Commission, numerous environmental groups and citizens were among those responding to SAVEs "call to action".

The next several years would entail negotiations between the Division of State Lands and Mr. Smith's legal team. After the completion of the negotiations, I was informed of the state's intention to file a legal claim of sovereignty on the 270 acres of sovereign lands located inside the 1973 dike.

Division of State Lands Recommends Quiet Title Suit

On 12 December 1995, before the Governor and Cabinet, the Division of State Lands (DSL) entered its recommendation to proceed with quiet title suit to reclaim sovereignty lands below the ordinary high water line of Smith's property. In its conclusions (Agenda item 16) DSL stated: *" the dikes in question are located below the OHWL of Lake Poinsett on state-owned sovereign land; Mr. Smith has not and cannot show any exclusive rights and privileges pursuant to which he is entitled to use these public trust lands; and appropriate legal action should be brought to have the dikes removed".*

State Files Suit against Developer

On 21 December 1995, in the Circuit Court for the Eighteenth Judicial Circuit In and For Brevard County, Council for the Board of Trustees of the Internal Improvement Trust fund, Mr. Jonathan A. Glogau, Assistant Attorney General for the state of Florida, filed suit against David A. Smith and Trustcorp of Florida (defendant) in a *"Complaint for Trespass Damages and Ejectment, with demand for a jury trial"*. The state's claim of sovereignty lands was now a matter of record.

Over the next several years—to 1999, a number of written legal arguments would be undertaken by the Plaintiff, Board of Trustees (Florida Governor and Cabinet), and the Defendant (David A. Smith). While these arguments and pleadings were significant, with the state defending its claim of sovereignty while Mr. Smith sought to counter such claims, the case would eventually be heard by the Eighteenth Judicial Circuit Court in Brevard County.

On 24 September 2001, in response to this writer's letter to Attorney General Butterworth, the Attorney General stated that depositions were scheduled for October; also, a hearing had been scheduled for November. The case was proceeding through legal channels available to both parties.

Attorney General Charlie Crist was elected to office in 2002. Former Attorney General Bob Butterworth left office due to the 8-year term limit. On 1 April 2003, in a letter to Attorney General Crist, this writer provided background information on SAVEs involvement with the Sabal Hammocks case. I requested Attorney General Crist provide SAVE with an update on the status of the Sabal Hammocks case.

On 26 June 2003, Jonathan A. Glogau, Chief, Complex Litigation for the Attorney General's office responded. Mr. Glogau informed this writer that "good faith mediation" was ongoing with Mr. Smith, with mediation to be completed by 25 July 2003. I was also informed that such mediation discussions were confidential by order of the court. Mr. Glogau stated however, that *"we are actively preparing for trial"*.

Trial Conducted in Brevard County

In January 2005, a five-day trail was held in the Eighteenth Judicial Circuit Court in Brevard County. Joyce and I attended every day of the trial. I was shocked to discover that the trial would not be decided by a jury, as previously demanded by the state. An agreement had been negotiated to accept the findings of Circuit Court Judge Bruce Jacobus. The state made a very strong case; one I was convinced was correct in their expert presentation.

As stated in a previous paragraph, all negotiations were, by law, confidential. After a long seven-month period, on 29 August 2005, in the final order written by Judge Jacobus, the Judge accepted Smith's argument that the OHWL was at the 12.1 ft. el, exactly where the illegal dike had been constructed in 1973. Read the extracted portions of the judge's final order in Chapter VII, Private Lands vs. Sovereign (Public) Lands. In the infinite wisdom of Judge Jacobus, the state may have lost the battle, but, in the infinite wisdom of this writer, I am convinced the "people" will win the war.

In order to avoid a potential precedent-setting situation, the state of Florida decided not to appeal the order of Judge Jacobus to the Florida State Supreme Court. At risk was the potential for development of other similar low-lying lands around the state, which is estimated to total some 500,000 acres should the Supreme Court uphold the Jacobus Final Order. However, the state won the last two cases that were presented to the State Supreme Court. Again, the writer is not an attorney. I will respect the decision of the Attorney General and his staff of attorneys.

On 20 September 2005, Florida Today published an interesting article regarding Mr. Smith and his legally challenged property. Complete with a picture of Mr. Smith, sitting at a picnic table at F. Burton Smith Regional Park (named after his father), Smith stated his willingness to "sell the land to the state as a conservation area". If the parties to a potential purchase of these lands fail to reach an agreement, in this writer's opinion, Mr. Smith will not be granted any permits to construct a large scale residential subdivision behind dikes, where 70 percent of the land is located in the annual to 10-year floodplain. Brevard County's Comprehensive Plan prohibits such development in the annual to 10-year floodplain of the St. Johns River.

The Brevard County Commission should never consider the top of Smith's illegal dike represents the 100-year floodplain, as Smith has stated in the past. One only has to look at New Orleans to understand this similar situation. On Smith's property, the dikes failed to contain heavy stormwaters (not a hurricane) in the fall of 2005. This writer received a call to visit the site and view the opening in the dike. The adjacent property, F. Burton Smith Park was flooded for over a month following the opening in Smith's dike.

Recently, this writer contacted the Brevard County Endangered Lands Committee, requesting the committee approach Mr. Smith for negotiating a fair price of the low-lying lands. Mr. Smith was mailed an application for submittal to the Endangered Lands Committee for evaluation. The Nature Conservancy and Florida Forever are a potential source of funding for the purchase. Together, we can claim a victory for the people, the St. Johns River, its fishery and wildlife.

As a part of any purchase agreement, the breeching or removal of the illegal dike should be the major action required to restore the area's floodplain to its natural state. In the final analysis of this 15-year process, this writer looks forward to the purchase of these marshlands and their return to the historic lakebed and floodplain of the St. Johns River.

More Fishing Trip Memories—My Best Day Catching Large Bass on the St. Johns River occurred on 26 May 1985. I lived on a canal leading into Lake Poinsett. After work on this memorable day, I decided to go fishing for a couple of hours before dark. From the boathouse, I lowered my boat into the canal water and moved slowly into the lake. In a spot where the water was approximately 6 feet deep, I began fishing with my favorite black & blue plastic worm. In less than 5 minutes I felt a light touch on the end of the line, which was stretching outward toward open water. The line tightened and I set the hook. One quick jump revealed a big bass was on the line. In short order, I caught and released three more bass, the smallest one being 5 pounds.

I was consumed with my fishing and unaware I had a spectator watching. A man called to me: *"Why are you putting the fish back in the water? If you catch another big one, would you give it to me; I would like to mount it and hang it above my fireplace—I will put your name on the plaque"*. Not really expecting to catch another big bass, I informed the gentleman if I caught another big one I would give it to him. I soon landed one that weighed 7 pounds. I quickly moved over to where he was located and gave him the bass. Later, just before dark, I had another very light touch on the line. The line tightened very quickly. I set the hook; almost immediately the bass wrapped around an object protruding out of the water. I proceeded to the object and had to take my rod and rotate it around the object to free the line. As I was freeing the line, the monster bass surfaced creating a huge splash. The reel I was using held 20-pound test line, otherwise the fish would have broken the line as it straightened out directly to the fish. I landed this beauty; it weighed 9 pound, 12 ounces. I was through fishing. My total catch included one at 5 pounds, one at six pound, 8 ounces, one at seven pounds, one at seven pound, 12 ounces and the biggest one at nine pound, 12 ounces. Total weight of the five fish was 36 pounds. All of these bass had completed their spawn cycle and were in a rare schooling situation. I mounted the nine pound, 12 ounce bass. It would be the last big bass I would ever mount. Six months later I formed the SAVE St. Johns River group; I was now a fully dedicated conservationist opposed to keeping the catch. Should I ever catch one over 11 pounds, I'll have a replica mounted for my collection.

In 1989, following my retirement from Martin-Marietta Corporation, I now had more time to fish. I set up a half-day guide trip with a father and his 7-year old daughter. We would be fishing in Lake Poinsett on the St. Johns River. This enthusiastic young girl informed this guide that she was going to catch a big bass. Her father informed me this was his daughter's first fishing trip. I wanted to give this young lady the best chance to catch a nice bass. I provided live bait (shiners) for this trip since she was not an expert in casting a rod and reel.

I anchored the boat near an island where the river current slowly moved through the area. I put a shiner on her hook and cast it out toward the island. In talking with her father, I informed him: *"She may not catch the first bass that bites, but I want her to land a bass without any help from either of us"*. In about 15 minutes, her bobber suddenly disappeared. The bass jumped once and the line went slack—the fish was gone. My young client had forgotten to "set the hook" before she started reeling in the fish.

I encouraged the 7-year old to be ready; she would have another chance to catch "a big fish". Again, I reminded her that when the "bobber goes under wait until you feel the line tighten, then jerk the rod and began reeling as fast as you can". Thirty minutes passed—no bite. This youngster was not discouraged. She was so attentive, watching the shiner pull the bobber near the grass line of the island. Suddenly, the bobber was moving faster—I told her to get ready, a bass was about to grab the shiner. The next thing I knew the bobber was gone; she remembered to wait until she felt the line tighten. She jerked that rod straight up and began reeling. Almost immediately, the "big fish" jumped completely out of the water. Still holding on and reeling, she was screaming with joy. *"It's a big fish"*. Indeed it was; over 6 pounds, this little girl was smiling as she admired her catch. I will never forget, not only the look on her face, but also that of her father. Upon our return to the boat ramp, Dad paid me for the trip and stated that it was the most satisfying $125.00 he had ever spent. Dad and his daughter were not the only ones satisfied; this guide was proud to have been a part of watching a child "catch a big fish".

CHAPTER VI—

RESTORING THE OCKLAWAHA RIVER

The beautiful Ocklawaha, a tributary of the St. Johns River, is one of the canopied rivers of Florida. The Ocklawaha was a corridor for fish and wildlife long before humans explored the region. These inviting waters provided prime habitat for many varieties of fish, including shad, striped bass, largemouth bass, as well as a variety of panfish.

Manatees routinely made their way up and down the Ocklawaha and the nearby St. Johns River. A mile-wide swamp/forest provided a refuge for the black bear, deer and other species of wildlife. These animals roamed freely along the shoreline of the St. Johns River to the Withlachocee Forest and Green Swamp in central Florida.

Man has not been nature's best friend along the Ocklawaha River. The area was well known for its beautiful and biologically rich eco-systems. Shortly after man arrived, the great cypress trees were cut and sold for their timber value. The landscape continued to be ravaged as man hunted, trapped and killed much of the wildlife. The earliest settlers hunted these animals for their hides and as a food source. Such activities were common. Fish and wildlife were abundant throughout the river and forest. Other "outdoorsmen" enjoyed great sport in bringing down a bear or a deer. As more and more settlers arrived, many wildlife species were facing an ever increasing struggle with man.

The free-flowing Ocklawaha managed to survive very well for centuries. Massive amounts of clean water from Silver Springs enabled the river to remain a picturesque and productive stream. The river remained in its natural state until the construction of the ill-advised Cross-Florida Barge Canal Project.

In 1826, the first government survey was authorized to determine the feasibility of excavating a shipping corridor across north Florida to simplify shipping between the east coast cities and the outposts along the Gulf of Mexico. Congress eventually authorized the Cross-Florida Barge Canal Project.

Construction Begins on Cross-Florida Barge Canal Project

In 1964, the U.S. Army Corps of Engineers (COE) started construction of the canal that would stretch across central Florida, from the Atlantic Ocean to the Gulf of Mexico. The width and depth of the canal would have to be sufficient to float large barges and boats. Rodman Dam and reservoir, complete with a lock-gated system was created as part of the Cross Florida Barge Canal in 1968. The results set in motion the destruction of the Ocklawaha and the loss of 9,000 acres of the river's floodplain. The Ocklawaha was now an obstructed river, and will remain as such until the dam is breeched.

The mistake in the construction of the Rodman Dam and reservoir was the result of poor judgment and greed by past political leaders. Special interest advocates decided nature could be altered to increase profits in the shipping industry. However, today the public, in general, overwhelmingly support the removal of Rodman Dam as evidenced in letters, speeches, meetings, etc. this writer supported for years. Previous Governors have supported the removal of Rodman Dam. Likewise, Governor Bush is on record as supporting the removal of the dam.

In 1969, the Florida Defenders of the Environment (FDE) was formed. Its purpose was to protect the natural resources of Florida. The original coalition consisted of a group of scientists, economists and lawyers who gave of their time and expertise to improve the policies and decisions affecting Florida's natural environment.

Throughout Florida's history, the Ocklawaha River, the largest tributary of the St. Johns River, has been regarded as one of the state's most beautiful waterways. Although the construction of Rodman Dam significantly impacted the Ocklawaha, the river remains a scenic and treasured asset.

FDE Halts Construction of Cross-Florida Barge Canal

The FDE is best known for its work in stopping the construction of the Cross-Florida Barge Canal.

The canal project was halted in 1971by President Richard Nixon. In 1990, President George Bush signed a bill de-authorizing construction of the canal. At that time, responsibility for the dam structures was transferred to the state of Florida.

FDE affiliate organizations now number over 50 local, state and national organizations. In the summer of 1991, this writer contacted Mr. David Godfrey, Project Coordinator for Restoration of the Ocklawaha. I requested FDE provide SAVE with the background data regarding restoring the Ocklawaha River. During the next two weeks, Mr. Godfrey and I discussed our mutual concerns for protecting the St. Johns River and its tributaries.

In a letter dated 26 August 1991 from Mr. Godfrey, I was informed that the FDE Executive Board had voted to support SAVEs effort to stop the proposed Sabal Hammocks Project (re: Chapter V). Subsequently, SAVE voted overwhelmingly to support FDE in pursuing the removal of Rodman Dam.

In a follow-up letter to the FDE on 16 September 1992, I provided a copy of SAVEs letter to Governor Lawton Chiles and the Cabinet, stating our support of FDE for the removal of the Rodman Dam. Governor Chiles and members of the Cabinet acknowledged SAVEs concern for removing Rodman Dam. Several cabinet members also pointed out their support for the Greenbelt Recreation and Conservation Area as prime habitat for various species of wildlife.

When the name Marjorie Harris Carr is mentioned, many memories quickly surface of this fine lady's commitment to preserving nature for future generations. Marjorie was very devoted to the preservation of "old Florida natural places". Her love and dedicated effort to protect the Ocklawaha River basin was clearly acknowledged when the Florida Greenway was designated: "The Marjorie Harris Carr Cross Florida Greenway". The 110-mile corridor traverses a wide variety of natural habitats and offers a variety of trails and recreation areas. The nature trail provides the means to explore old Florida for all who visit the area.

A Call to Remove Rodman Dam

Marjorie Harris Carr, John H. Kaufmann, Ph.D., and David Godfrey, FDE Project Coordinator, documented their concerns in a paper entitled: *"Cross Florida Greenway Plan Should Include Restoring Ocklawaha River"*, dated August 1992 and published by the FDE. A reprint of the publication follows:

Rodman Dam located on the Ocklawaha River, southwest of Palatka

"Rodman Reservoir should be drained immediately so the Ocklawaha River and its floodplain forest can begin the natural process of restoration. Rodman should have been drained when the Cross Florida Barge Canal was halted in 1971, and certainly should now that the former canal lands will become the Cross State Greenway Recreation and Conservation Area. That the Ocklawaha River is still in bondage in 1992 is an environmental and political scandal of classic proportions.

From 1974 to 1976, the Corps of Engineers, and other federal and state agencies, conducted the $2.5 million Cross Florida Barge Canal Restudy. This massive research effort answered the very questions being asked today regarding the economic and environmental pros and cons of restoring the Ocklawaha River. The results of this exhaustive study showed that the canal project should be stopped and the Ocklawaha River should be restored. Florida's Governor and Cabinet agreed with the findings and the Florida Legislature passed a bill supporting their position.

In 1977, the Corps of Engineers, the U.S. Forest Service and Florida's Department of Natural Resources prepared the plan, "Alternatives for Restoring the Ocklawaha River..." Restoration was set to begin, but intransigent canal supporters kept Congress from reauthorizing the canal project until 1990, preventing the restoration of the river until now.

The information needed to again decide whether or not to restore the Ocklawaha River is already available from the earlier studies and from information gathered for the current Canal Lands Advisory Committee. Calls for unnecessary additional studies would delay indefinitely—at additional cost to taxpayers—a decision for action that is already long overdue.

The expensive effort to maintain Rodman as a fishery has lasted over two decades at a cost approaching $20 million. The Corps of Engineers has brought its considerable expertise to the problems at Rodman to no avail. Part of their difficulty lies in the fact that Rodman is not a true "reservoir" but merely a shallow backwater spread over the original floodplain of the Ocklawaha River. The Corps has commissioned extensive water weed studies and tried several different control regimes, but Rodman remains shallow, weed-filled, subject to massive fish kills and beset with submerged logs.

Recreation at Rodman is strictly limited to fishing and seasonal duck hunting. Dense swamp vegetation in the upper Rodman Pool, and dense water weeds in the lower pool, renders much of the 9,000 acres inaccessible. Stumps, floating logs and water weeds prohibit the swimming, sailing, water skiing and recreational power boating that most people expect to enjoy on a lake. Thus, 16 miles of the proposed Cross Florida Greenway is currently sacrificed to provide fishing in a limited area to a relatively small number of reservoir fishing enthusiasts. The restored river and forest, however, will provide river fishing, canoeing, camping, hunting, hiking, swimming and scenic boat tours.

Rodman's continued existence also conflicts with current federal and state statutes. The Federal Clean Water Act and its accompanying regulations mandate that balanced biological communities of native plants and animals be maintained. Rodman does not meet this test. The restored river and swamp forest would. The State Comprehensive Plan calls for the protection and restoration of natural water systems in lieu of structural alternatives (including dams), and for the maintenance of ecologically intact systems.

A free-flowing, self-maintaining Ocklawaha River will be the most unique and beautiful section of the Cross Florida Greenway, and it will be of the most value to the public. This will, of course, translate into economic advantage for the adjoining municipalities. Ecotourism, which is just getting underway in Florida, promises to be a substantial economic factor in the future that should be carefully considered in planning the Greenway.

The Greenway is an extremely narrow belt running across the State. This narrow corridor will have to be rigorously protected from enterprises that would overwhelm other nature-based recreation resources. The Greenway must be developed for the general public, rather than for special recreational interests.

Already we are seeing the potential threat of the Greenway being divided up into separate sections dedicated to use by particular interests. On the west end, a giant marina could dominate use of the barge canal with boat traffic. In the middle section of the Greenway, a proposed horse park, including polo fields, might result in a Greenway overwhelmed by intense equestrian activities. And, of course, the continued maintenance of Rodman Reservoir would mostly limit access to a small specialized group of fishermen.

The overall public interest demands that Rodman Pool should be immediately drained and that the Ocklawaha River and its floodplain forest should be restored as an important part of the entire Cross Florida Greenway".

In 1993, Governor Chiles and the Cabinet voted unanimously to have the Rodman Dam removed. Following the vote, months passed with little media or continued public support to obtain the necessary funding. As support slowly faded, the issue also faded in the halls of the state legislators.

Fishing Open Waters of St. Johns River vs. Rodman Reservoir

On April 28, 1994, Joyce and I fished a bass tournament on the St. Johns River out of Crystal Cove Marina, in the Palatka area. We were members of a three state (Georgia, Florida and Alabama) bass tournament organization known as "Guys and Dolls Tournament Trail". We fished Rodman Reservoir on two previous trips to the area to "practice" for the upcoming tournament. We caught and released a few bass during two trips to the reservoir. When comparing the number of hours we fished vs. the number of bass caught, the average was one bass for every 3 hours both of us fished. The largest bass we caught weighed 4 pounds.

The next day we made one "practice" trip on the St. Johns to nearby Crescent Lake. We caught and released five bass in less than two hours. On the day of the tournament, we caught 13 bass, keeping the largest five fish for weigh-in and releasing the other 8 back into Crescent Lake, before departing for the weigh-in site. Our tournament limit of five bass weighed almost 14 pounds.

With 90 boats entered in the tournament, we missed finishing in first place by one pound. A few of the 90 boats fished Rodman Reservoir. We made the right decision; we ignored the entrance to Rodman and traveled through Dunn's Creek to reach Crescent Lake. We caught more and bigger bass in Crescent Lake than the anglers who fished Rodman Reservoir. We placed a beautiful plaque on our "trophy wall" in our home. Our cash award was probably spent fishing our next scheduled tournament on Lake Kissimmee.

Rodman Reservoir Designed as Transportation Corridor, not a Fishery

As in most newly impounded water bodies, during the first several years, rich nutrient levels and "new" shallow water shorelines result in excellent fishing, as was the situation at Rodman Pool. Eventually, catching a lot of bass in a reservoir such as Rodman will become more difficult. Fishing pressure alone accounts for much of the decline in quality fishing. Water current in natural rivers keep sediments from building up on the bottom. In small reservoirs such as Rodman, the function of current is significantly reduced (the dam stops current flow). Also, sediment build-up continues to cover the reservoir's bottom.

In a short period of time following major storm events, fish kills can occur as a result of insufficient levels of dissolved oxygen. The levels of dissolved oxygen drop, sometime significantly, as a result of the "stirring up" of accumulated sediment locked in by the dam. Add to this condition, the higher water temperatures due to the absence of current, coupled with exotic plants such as hydrilla which reduces oxygen in the water. Often times, cloudy days following a storm event contribute to fish kills in a reservoir type environment due to the prolonged absence of sunlight which reduces the generation of oxygen in the water.

As a St. Johns River fishing guide for a number of years, I preferred to be "free" to explore open waters of the river and its many connecting lakes than to be restricted to the confines of a reservoir such as Rodman. On each fishing trip with a client, I provided the angler my business card before we departed the boat ramp. On the card, in large print was the statement: No Fish—No Pay. I never experienced a trip that the client did not catch fish. All my guide trips were on the open waters of the St. Johns River.

As briefly stated in an earlier chapter, I have been bass fishing on every lake throughout the St. Johns River, from Lake Hell N' Blazes in Brevard County to Green Cove Springs, near Jacksonville. Joyce enjoys bass fishing and is quite an angler. In addition to fishing the St. Johns, our travels include the Stick Marsh (Upper Basin of the St. Johns), the Harris Chain of Lakes, the Kissimmee Chain of Lakes, Lake Okeechobee, and Lake Fork in Texas. The natural setting of these magnificent water bodies frequently call to the spirit within both of us.

In the 1970s, I especially enjoyed fishing Lake George, Lake Dexter,

Lake Woodruff, and Crescent Lake. I have memories of catching and releasing many bass on these lakes. I spent a number of week-ends at Ormond's Jungle Den (now named Blair's Jungle Den) on the river at Astor. I recall a trip when I was wade-fishing the shoreline near the Silver Glen Springs outlet into the lake.

The area was great for bass fishing. After stepping in a hole or two, I looked down to view the bottom. I noticed large and round depressions in the sandy bottom. There were many of them. After remaining still for a few minutes in the clear water, I soon recognized the fish "spawning" in the many "beds" were tasty shell-crackers, a member of the "bream" species. I stopped bass fishing and purchased two boxes of live wigglers. I returned to the spot using light spinning tackle and caught 31shell-crackers before I used up all my wigglers. That evening, my family enjoyed a great meal of fresh fish.

Beautiful and popular Salt Springs on northwest shore of Lake George

On some trips to Lake George, I fished near the structures located at the St. Johns River entrance into the south end of the lake. On other trips, it was fun to travel in the opposite direction to other nearby lakes, such as Lake Woodruff and Lake Dexter. These lakes also provided great fishing, in addition to an enjoyable boat ride. Joyce and I enjoy fishing the larger natural lakes and river areas of the St. Johns River. Boaters can travel endlessly on these waters, and seldom need to fish a particular "hole", as is the situation in Rodman Reservoir. On the few occasions I fished Rodman, if the trolling motor wasn't bouncing off stumps, I was losing "fishing" time cleaning hydrilla and other exotic vegetation from my trolling motor.

On one trip to the Lake George area in 1963, our family vacationed at Hall's Lodge, in Astor. My 11-year old daughter (Donna) dropped her hook and line into the water's edge in front of the lodge to measure the depth of the water. This is sometimes necessary to adjust the bobber's setting. Donna's bobber quickly disappeared. She pulled out a two-pound bass that struck her gold-colored empty hook. To this day, Donna speaks of her big bass caught on an empty hook.

During this same trip, this writer hired a guide to take me bass fishing (my boat was in the repair shop). After several hours of fishing in Lake Dexter, having caught and released six bass, we returned to the lodge for lunch. There were some storm clouds gathering. My guide asked if I wanted to catch some large bedding bream (blue-gills). He stated that plenty of these fish were located near the lodge. Using light tackle, these fish feel like two pound bass. Considering the weather conditions, I agreed to fish for blue-gill. We backed out of the dock at the rear of the lodge and stopped the boat just 40 feet from the lodge. I caught about 50 of the spawning blue-gills in about two hours.

Our family enjoyed the Lake George area so much that I bought property off SR-19 south of Salt Springs and placed a mobile home on the site. We returned numerous times to enjoy the nearby recreation opportunities of Salt Springs, Alexander Springs, Silver Glen Springs and fishing the lakes in the area.

This writer included the above personal stories of my experiences in the area to make a point. Restore the Ocklawaha River; in its natural state, the river will be utilized by more families and tourists. Restoring this river will also expand the area's economy.

Federal Government Supports Restoration

In 1997, the U.S. Environmental Protection Agency (EPA) provided Florida $640,000 dollars to create a wetlands restoration project by getting permits for restoration of the Rodman Reservoir. The federal agencies again offered their assistance in the form of technical resources and to the extent possible, financial resources. The EPA reiterated the fact that there has always been agreement on the basic philosophy of restoration. A spirit of cooperation existed between all levels of government—local, state and federal agencies.

In a letter from FDE President Alyson C. Flournoy, dated 25 November 1997, SAVE was requested to offer our voice and support as a member of an ad-hoc Alliance to Restore the Ocklawaha. SAVE supported the FDE in this effort to remind state legislators that removing Rodman Dam was not a local issue, but indeed was a state-wide issue of importance. State Senator George Kirkpatrick took on the Rodman Dam issue as a pet project to prevent the removal of the dam.

SAVE Supports Restoration

On 19 January 1998, I wrote a letter on behalf of SAVE to Senator Jack Latvala, Chairman of the Florida Natural Resources Committee. In my letter, this writer also spoke as Central Florida Regional Director for the Florida Wildlife Federation. My letter addressed the need for action by the Natural Resources Committee to sponsor funding in support of the state's previous approval to restore the Ocklawaha by removing Rodman Dam. I informed the Chairman that I was speaking for thousands of members of the Florida Wildlife Federation and supporters of SAVE. I provided "info" copies to all Natural Resources Committee members, including Senator Kirkpatrick.

In response to a letter I wrote to Governor Jeb Bush in April 1999, the Governor responded in part as follows: *"The debate about restoring the free-flowing river or retaining Rodman reservoir is again heating up...Since this appears to be an environmental and economic policy question charged with emotion, there are several guiding principles I will use to help reach a better understanding of restoration requirements and the importance of the Rodman reservoir to the region...*

First I strongly believe that official state policy on the Rodman Dam should be based on objective scientific information, not politics. Second, I believe that Florida's policy on the Ocklawaha/Rodman issue should be based on the most responsible management of our limited resources...finally, our policy should ensure that environmental protection needs are balanced with the livelihoods and quality of life for families of limited means. Rest assured that I will objectively consider all sides of this issue before making any decisions on this matter".

In a letter from State Treasurer Bill Nelson dated 16 April 1999, Mr. Nelson assured this writer that *"As you know, my record on this matter is clear and I remain committed to the restoration of the Ocklawaha River".*

FDE Alliance to Restore the Ocklawaha River Initiates Action

FDE wrote a two-page letter to Governor Bush in January 2000.

Forty-five signatures representing 45 separate membership organizations who comprise the "Alliance to Restore the Ocklawaha River" on behalf of the FDE was a significant matter. The letter contained three additional pages of the 45 organizations that signed their names and positions. The final page was a listing of local, state and federal agencies that were provided a copy of the letter.

SAVE was one of the 45 signers. The letter left no doubt that the restoration of the Ocklawaha River had wide-based community, state and national support. The letter addressed many of the statements already discussed in this chapter. The letter contained convincing evidence. Governor Bush was reminded that the previous four Governors of Florida had supported the restoration. The Governor responded to the FDE with assurance that he supported the restoration of the Ocklawaha River, including the removal of Rodman Dam.

FDE Executive Director Leslie E. Straub wrote Governor Bush on 17 July 2000, expressing the FDEs gratitude for his recent decision to move forward with the restoration of the Ocklawaha River. FDEs letter further pointed out: *"the 30-year disruption in the life of this unique and beautiful river will quickly be erased as it returns to the healthy, dynamic ecosystem it once was"*.

A view of dead trees in Rodman Pool; flooded and destroyed by construction of Rodman Dam

Controversy over Rodman Dam Continues

In 2000, Senator George Kirkpatrick began his last legislative session to obstruct the restoration of the dam that now bears his name. In the House of Representatives, a bill was filed by Representative Fuller, (HB1599). In the Senate, a similar bill was filed by Senator King (SB1976). Both bills would designate Rodman Pool as a state recreation area managed by the Florida Department of Environmental Protection's Division of Recreation and Parks.

Any changes to the dam, spillway or navigation locks would be prohibited without specific legislative approval. These conditions would assure the return of this contentious issue to the next legislative session and give Senator Kirkpatrick's allies a better opportunity to kill funding for the restoration. These bills would allow restoration opponents to continue confusing the public by attacking restoration plans as the destruction of a designated state recreation area.

In addition, these bills would authorize the DEP to trade land for those portions of the Ocala National Forest that are now illegally impounded or occupied by the dam. Senator Kirkpatrick perhaps could now work a deal with the U.S. Forest Service in a future federal administration to save the dam. Finally, the bills directed DEP to determine how to compensate the owners of private land submerged by the impoundment, whose monetary claims are mounting because of the state's delay in removing the illegal flooding of their property.

These bills troubled FDE; their position was to reject these bills—the Kirkpatrick (Rodman) Dam, reservoir, and the Buckman Lock are unworthy of inclusion in Florida's system of state parks and recreation areas. Maintenance alone would cost at least $300,000 to $500,000 per year, excluding the cost and /or repair or replacement of the structures.

On 3 April 2000, the DEP ordered Buckman Lock at Rodman Reservoir to be closed indefinitely. Manatee protection devices were ordered to be installed to prevent further death or harm to the federally protected manatees. Since 1977, at least 10, but perhaps up to 20 manatees have been killed by Rodman Reservoir's water control structures.

The 2000 legislative session ended without the creation of a legislatively-managed Rodman Reservoir State Park. This came as no real surprise to many FDE supporters.

On 27 May 2000, the Florida DEP submitted information for the Environmental Resources Permit needed from the SJRWMD for restoring the Ocklawaha by breaching Rodman Dam. FDE looked forward to receiving the scientific, ecological and economic data. FDE felt confident, as all previous studies have concluded, breaching Rodman Dam would greatly benefit both the St. Johns River and the Ocklawaha river watershed.

The afternoon of 14 July 2000, Florida's office of the Governor began phoning restoration supporters around the state to tell them Governor Bush has decided to support restoring the Ocklawaha River. That same day the DEP submitted their long-term plan and Environmental Impact Statement for breaching the dam, to the U.S. Forest Service. Manley Fuller, President of Florida Wildlife Federation stated: *"Sportsmen of Florida applaud the Governor's decision. Restoration of the Ocklawaha is important for the long-term health of the whole St. Johns River watershed".*

In late September and early October 2000, another major fish kill of approximately two million fish occurred in Rodman Reservoir. Other major fish kills occurred in 1985 and 1988. According the Florida Fish and Wildlife Conservation Commission (FWC), the two previous fish kills were even larger than the 2000 event.

In 1985, over 2,500,000 fish died; in 1988, there were 8,500,000 dead fish. Reasons for the fish kills were similar; heavy storm events disturbed sediments on the bottom of the reservoir, these storms followed major drought conditions in the area. Uninhibited growth of exotic and invasive hydrilla as well as heavy rains brought dark stained water, low in dissolved oxygen into the reservoir from the surrounding wetlands. **A free-flowing Ocklawaha would not have experienced such massive fish kills.**

The Governor's Fiscal year 2001 budget included funding to complete the restoration engineering plan for the Rodman Reservoir. Governor Bush presented his conservation agenda at the Everglades Coalition meeting in January 2001. He proudly displayed a "Free the Ocklawaha River" button attached to his shirt pocket. Many groups were relieved that after 32 years of blocking the river's natural ability to sustain itself, restoration seemed most likely to get underway.

Two months later, on 1 March 2001, Senate Majority Leader Senator King presented SB 1246 accompanied by House Bill 1085, sponsored by Representative Pickens (Palatka) to preserve Rodman Dam. The "Rodman Dam Preservation Bill" would take the Rodman Reservoir portion of the Cross Florida Greenway out of direct management by DEP by requiring decisions to be approved by the legislature. This anti-restoration legislation would create a conflict between state and federal laws by authorizing non-federal facilities on federal lands; moreover, a terrible precedent would be set by requiring legislative approval of management decisions for a park. Governor Bush threatened to veto both bills.

Meanwhile, Senators King and Horne, members of the Senate Appropriations Committee, countered the Governor's request for restoration funding with $1.6 million dollars for three facilities at Rodman Reservoir. The committee designated $300,000 dollars for a Rodman Park, $500,000 dollars for boat ramps and parking, and $800,000 dollars for a fish hatchery to be located at the dam. The House Appropriation Committee kept the $800,000 dollars for restoration of the Ocklawaha River in its version of the state budget.

When the Rodman Bill was presented in the Senate Natural Resources Committee at mid-session, it was replaced with a version of the bill with neutral language which did not prevent restoration. However, the core of the bill was now to provide money for a user's facility on the Rodman Reservoir and money for the state to study water quality in the Ocklawaha River basin. The bill also required construction of small spectator bleachers, paved parking and a pavilion adjacent to the reservoir. The facility money was an obvious effort by reservoir supporters to increase the value and use of the reservoir by using state funds to subsidize the bass tournament business of a prominent restoration opponent.

An unsuccessful effort to forge a compromise in the state legislature resulted in no funding for a water quality study required for restoration of the Ocklawaha River and for a bass tournament facility. On 15 June 2001, Governor Bush vetoed facility improvements to accommodate bass fishing tournaments at Rodman Reservoir. Bottom line: no restoration funding, no water quality study by SJRWMD, and no Rodman Reservoir bass tournament facility.

Having observed state legislative methodology for the past 20 years, this writer agrees that legislators should listen to the voice of their constituents; the people who put them in office. In my opinion, Senator King, Senator Kirkpatrick, Rep. Albright and Rep. Putnam acted to respect the desires of their constituency. However, there are times when the intellect of responsible legislators should rule over the desires of a lesser informed constituency.

The removal of Rodman Dam is such an "intellect calling". A free-flowing Ocklawaha River will not displace any resident citizens of a good "fishing area". In fact, this entire area once was known as the "Bass Capital of the World". I am aware that much of the drive to keep Rodman Reservoir as a "locked body of water" comes from vocal bass clubs. Their view is wrong. As a bass fisherman myself, it is difficult to communicate the environmental rewards to be realized to these "pros", without perceived damage to their "fishing hole".

Personal Experience with a "Weir"

In Brevard County, Lake Washington is the third lake on the St. Johns River as the river flows north. Many years ago, this writer opposed the construction of a "weir" at the lake's exit point on the north end of the lake. The weir was considered necessary to create a potable water supply for the booming population of south Brevard County. All boaters have lost direct access from Lake Washington to the river north of the weir. Melbourne area residents must trailer their boats a minimum of 18 miles via I-95 to Cocoa, then proceed 5 miles west to the nearest public boat ramp. The same route applies to residents desiring to fish the three lakes south of Cocoa; only difference is the travel is reversed.

To my knowledge, no other weirs or dams are planned for the St. Johns River, although plans are being discussed for the extraction of surface waters from the St. Johns to support even more increases in Florida's population. As a potable water supply, it is expensive to treat the surface water of Lake Washington; not to mention the almost drought conditions on the river north of the weir anytime a shortfall of rain occurs. The first priority is to refill Lake Washington.

As this book goes to press, only canoes and airboats can navigate the first 70 miles of the St. Johns River. The weir captures most of the rain south of Lake Washington, releasing minimal flows northward. After 40 years, I have come to "live with it". Often times, Joyce and I trailer our boat to other area lakes, in search of enough water to float the boat.

Rodman Dam Saga Continues

The year 2002 would continue to provide conflicting issues regarding the restoration of the Ocklawaha River. In early 2002, the U.S. Forest Service (USFS) notified the state of Florida that Rodman Dam's impact upon portions of the Ocala National Forest necessitated the need for the state of Florida to remove Rodman Dam and begin restoring the Ocklawaha River. Such action by the state would, in turn, restore USFS federal lands.

A time-table was established; restoration would result in a free-flowing Ocklawaha River by June 2006. Permitting would be necessary for water quality purposes. Starting in 2003, the reservoir would begin gradually draining the impoundment. Finally, 2,000 feet of the dam would be removed, allowing the river to flow freely into its historic channel.

In January 2002, yet another Final Environmental Impact Statement (FEIS) for restoration was released. Like other environmental studies before, the FEIS documents a comprehensive evaluation of the environmental and socio-economic impacts of removing Rodman Dam. Restoration would include restoring 16 miles of the Ocklawaha and more than 7,000 acres of valuable floodplain forest, much of which is already in public ownership.

If restoration opponents manage to stall the restoration project, the U.S. Forest Service publicly stated that it will consider the state to be trespassing on federal lands. The federal government could take over the project and hold the state accountable for the entire cost.

SAVE Rodman Reservoir, Inc. filed an appeal against the U.S. Forest Service Final Environmental Impact Statement and Record of Decision which requires the restoration of the Ocklawaha River by June 2006. On 8 March 2002, the Alliance to Restore the Ocklawaha River responded to the Save Rodman Reservoir, Inc. latest efforts to stop the restoration project. The Alliance, led by FDE filed "interested party comments" supporting the earlier U. S. Forest Service decision. The Alliance rebuttal systematically and unequivocally rejects Save Rodman Reservoir, Inc. appeal as lacking merit.

On 19 July 2002, the Florida Department of Environmental Protection declined to sign the special use permit that allows the state-owned portions of Rodman Dam and Rodman Reservoir to continue to flood the Ocala National Forest. The U.S. Forest Service is now in charge of managing this land and part of the dam. The U.S. Forest Service has been consistent in supporting the removal of Rodman Dam since the 1970s. Over a third of the dam is in the Ocala Forest and approximately one square mile of the reservoir is submerged beneath the waters of the reservoir.

FDE considers transferring responsibility for restoration to the U.S. Forest Service to be a positive step, since funding the restoration by the state of Florida has not been successful. Also, FDE feels the federal government generally has deeper pockets. Securing funding from the Florida Legislature is unlikely, at least in the near future.

On 14 July 2003, Governor Bush vetoed the George Kirkpatrick Reserve Bill. Governor Bush continues to remain consistent in his desire to ensure restoration of the Ocklawaha River and the removal of Kirkpatrick (Rodman) Dam. In a letter to Governor Bush dated 8 July 2003, Attorney General Charlie Crist asserted his support for restoring the Ocklawaha River. Former Attorney General Bob Butterworth also supported Governor Bush's position of returning the Ocklawaha to a natural free-flowing river.

Bass Tournament Surveys Support FDEs Position to Remove Rodman Dam

Earlier in this chapter, I spoke of my earlier experience fishing the Rodman Reservoir. I pointed out that Joyce and I experienced more successful fishing by visiting the areas abundant lakes on the St. Johns River. Our bass fishing experiences on Rodman Reservoir as opposed to the nearby waters of the St. Johns River was borne out in an FDE study of bass fishing tournaments between February 2001 and August 2004.

The source of the FDE study was the records of bass tournaments held in Putman County and published in the Palatka Daily News. Anglers won more tournaments with bigger fish on the free-flowing St. Johns River's lakes, especially Lake George and Crescent Lake than in Rodman Reservoir. The St. Johns River and its natural lakes in the surrounding area are the reason the area has been historically known as the "Bass Fishing Capital of the World".

Nick Williams, Executive Director of FDE, rightly stated: *"there are many excellent opportunities in the area and the taxpayers of the state should not have to subsidize the reservoir"*. The study of fishing tournaments contradicts a claim that Rodman reservoir holds the biggest fish in the area. The study revealed such claims were not accurate.

As reported by the Palatka Daily News, the average weight of big bass caught on Rodman Reservoir (6.88 pounds) and in the St. Johns River system of lakes (7.04) was close; the results were comparable. Who would argue which big bass was bigger, the 6.88 pounder, or the 7.04 pounder. If I caught the 7.04 pounder, I would certainly argue: "My bass is bigger than your bass".

Of all bass weighing more than 10 pounds, seven were caught in the river (including the largest in the study—an 11.84 pound bass caught in Dunn's Creek) while just two were caught in the reservoir. Knowing anglers as I do (I am an angler too), do not try to convince the anglers who caught the two bass over 10 pounds at Rodman Reservoir that the St. Johns River and its lakes are a better place to land "that big one". The same holds true of the anglers who caught the seven bass over 10 pounds in the St. Johns River and its lakes. These anglers will remain committed to fishing "the big waters for big bass".

Two of the area's major tournaments, the Wolfson Children's Hospital benefit (which begins on the St. Johns River) and the Save Rodman Reservoir benefit (on Rodman Reservoir) were compared to get a better glimpse of the total weight of catches on the river vs. the reservoir. Both tournaments had a seven-fish limit. The average winning catch weight of river bass caught at the Wolfson benefit was 31.16 pounds. The average winning catch weight of reservoir bass caught at the Save Rodman Reservoir benefit was only 23.51pounds. If one does the math, the St. Johns River system provided 25 percent more weight than those who fished the Rodman Reservoir.

The River City Tournament series was also reviewed to determine if there was a difference between bass caught in the St. Johns River system vs. the Rodman Reservoir. The River City Tournament series is a monthly series, using an 8-fish limit, with two divisions, one division fish the Rodman Reservoir and the other division fish the St. Johns River system. While big bass weighed in from both locations were about the same, those anglers fishing the St. Johns River system won the total weight category by over 4 pounds per tournament.

If this study was used as evidence in a legal setting, and if I was defending the St. Johns River advocates, I would inform the Judge (he or she) that: "The St. Johns River advocates rests its case".

Conservation Groups Consider Legal Action

On the subject of legal action, in February 2006, a coalition of conservation groups filed a 60-day "Notice of Intent to Sue" in federal court on grounds that the U. S. Forest Service has violated the Endangered Species Act by not pursuing the restoration of the Ocklawaha River. The Endangered Species Act requires a 60-day notice prior to allowing a lawsuit to be filed.

The conservation groups stated that unless the U.S. Forest Service consults with the U.S. Fish and Wildlife Service and the National Marine Fisheries Service, legal action will be initiated. The coalition of conservation groups is represented by Wild Law attorney Brett Paben; the groups include Save Our Big Scrub, Florida Defenders of the Environment, Putman County Environmental Council, Marion County Audubon Society, Save the Manatee Club, Defenders of Wildlife and Wild South.

The suit is part of an effort to move forward with restoration of the Ocklawaha River. The suit seeks protection of the Florida manatee and short-nose sturgeon being harmed by the maintenance of Rodman Reservoir and Rodman Dam. The suit further claims that for the past three years the dam has been illegally occupying federal lands in the Ocala National Forest. During this 60-day Notice, the U.S. Forest Service has, in fact, been consulting with the U.S. Fish and Wildlife Service and the National Marine Fisheries Service. It is the desire of the FDE and its supporters, that the issue be resolved without necessity of time and expense of proceeding with the lawsuit.

Federal agencies should avoid a potential lawsuit; act responsibly and restore the Ocklawaha River. It is past time to return the historic Ocklawaha to a free-flowing river.

More Fishing Trip Memories—In May 1989, I was approached by a resident of Cocoa Beach to discuss one or more guided trips on the Upper Basin of the St. Johns River. In his mid-thirties, a jubilant man with an outgoing personality, he proceeded directly to his point for the "one or more" fishing trips to the river. To protect his privacy, I will call him "Dan".

Dan wanted to schedule one trip a week for as long as it would take him to land a 7-pound bass. His preferred day for fishing was Wednesday. I could not guarantee he would land a 7-pound bass, although he assured this guide he knew bass fishing very well. I agreed to schedule him to fish on Wednesday of each week. Fishing was generally good each Wednesday we were on the water. Three weeks passed; no 7-pound bass. We had some conversations while fishing; I was somewhat interested to know his profession. I never found out what he did for a living. He freely spoke of his wife's profession—she was a "entertainment dancer". I did not pursue that conversation any further.

On the morning of the fourth Wednesday, I stopped my boat near a point on the west side of Lake Poinsett; I knew bass were in the area. In less than 30 minutes after moving slowly to the very outside point of some vegetation, Dan hooked a large bass. Although the bass never came completely out of the water as it attempted to free itself from Dan's lure, I saw the open mouth of a big bass. A couple of minutes later, Dan landed his fish. Using digital scales I kept in the boat, his catch weighed 7 pounds, 4ounces. When we arrived back at the boat ramp, Joyce met us. Dan proudly showed off his catch. Joyce said to me: *"That would have been my catch if you were not on a guide trip today"*. Dan released his catch after posing for a picture with it. I never saw or heard from Dan again.

CHAPTER VII—

PRIVATE LANDS VS. SOVEREIGN (PUBLIC) LANDS

The ownership of thousands of acres of submerged lands bordering Florida's rivers, lakes, and bays has been debated and eventually resolved by a court of law. Numerous times, ownership disputes are not challenged as properties are developed on portions of sovereign state-owned lands. Planning and growth management agencies representing local governments have only limited knowledge or expertise in matters of state sovereignty land laws, the state constitution, or public doctrine as they relate to this complicated issue.

In 1982, I built a home on a canal bordering Lake Poinsett in Brevard County. In 1994, a state Ordinary High Water Line (OHWL) survey conducted as part of SAVEs legal challenge to the proposed Sabal Hammocks project revealed my prior residence was potentially located below the OHWL. In the late 1950s and early 1960s, Lake Poinsett's marshlands were dredged (up to 18 feet deep); the spoil was stacked to meet the 100-year floodplain elevation. Over 75 homes were built on the dredged property. The state was never aware of the development until most of the homes were already built. However, the Florida Division of State Lands had no interest in reclaiming these lands since the residential development was over 90 percent "built out" at the time of the survey.

Florida Law Review, Univ. of Florida College of Law, 1990, Richard Hamann & Jeff Wade

Under the public trust doctrine, the State of Florida gained title to the beds of navigable lakes, streams, and tidal waters upon admission to the Union in 1845. These publicly owned submerged lands are perhaps the

View of St. Johns River looking south from SR 520

state's most valuable natural asset. Originally valued primarily as avenues for trade and travel, water supply sources, and conduits for the disposal of wastes, Florida's waterways are now also appreciated for their ecological, recreational, and aesthetic values. Floridians and visitors flock to enjoy the river, lakes, bays, and shores of the Sunshine State. Submerged lands, however, comprise some of the most sensitive environments in the state and provide essential habitat for endangered species and other wildlife.

Recreational and commercial fisheries depend upon maintaining the integrity of submerged lands. If wetlands are the "soul" of Florida, then submerged lands are surely its lifeblood. In Florida's earliest environmental battles, advocates fought to protect these aquatic environments from privatization and wholesale destruction. Strong judicial support for the public trust doctrine averted several attempts to divest the public of this resource.

As public support for environmental protection has grown, elected officials have steadily increased the protection given to publicly owned submerged (sovereign) lands. The 1968 Constitution codified the public trust doctrine and limited the government's power to dispose of these lands. The title to all submerged lands is now vested in the Governor and Cabinet, as Trustees of the Internal Improvement Trust Fund. It may be surprising to find, however, that the extent of public ownership of submerged lands remains a hotly contested issue. In the 1970s, private landowners began asserting ownership of submerged lands. There have been many lawsuits filed, both by upland landowners, as well as conservation-oriented organizations. The state has successfully defended ownership in a number of court cases around the state of Florida.

The Florida Supreme Court ruled in favor of the state in the Coastal Petroleum Co. vs. American Cyanamid Co. (1986). The state and Mobil Oil Corporation thus proceeded to litigate the OHWL issue. On the eve of the trial, however, the parties reached a settlement. The phosphate companies agreed to give the state all of the land it claimed, additional land in the 25-year floodplain, and several thousand acres of mined lands for park development. After spending 7.65 million dollars on the litigation, the state had achieved its goals on the Peace and Alafia Rivers, but had not established new precedent on the issue of how to determine the OHWL.

Public Trust: The Basis for State Ownership—States derive their ownership of the lands under navigable waters from the English common law doctrine by which the sovereign held title to the beds of navigable and tidal waters are held in trust for the people. After the American Revolution, each state acquired title to the lands beneath its navigable waters under the public trust doctrine. New states joined the union on an equal footing with the original thirteen states, including *"the right and duty of the states to own and hold the lands under navigable waters for the benefit of the people"*.

In 1819, the United States acquired the territory known as East and West Florida from Spain under the Treaty of Cession. When the United States took possession of the area two years later, the public trust doctrine required the government to hold the lands under the navigable waters, including the shores of spaces between the OHWL and low water marks and tidelands, for the use and benefit of the state that was to be subsequently formed.

Florida became a state on 3 March 1845, and under the equal footing doctrine, took title to all sovereignty lands within its jurisdiction, except those parcels previously granted to private interests by the Spanish government, or conveyed out by the federal government while Florida was still a territory. The sovereign's right of title included the responsibility to hold the lands in the public trust for the benefit of the people, with restrictions on alienation and use of the lands.

The doctrine has been incorporated in Article X, Section 11 of the Florida Constitution, which states that sovereignty lands are *"held by the state...in trust for all the people"* and that *"sale of such lands may be authorized by law, but only when in the public interest"*.

Navigability—Only nontidal waters that are navigable are subject to state ownership under the public trust doctrine. In general, bodies of water that at the time of statehood in 1845 were used or capable of being used in their ordinary and natural condition for trade or travel by the means common in the local area for waterborne transportation, are deemed navigable. *"A stream of sufficient capacity and volume of water to float to market the products of the country will answer the conditions of navigability,... whatever the character of the product, or the kind of floatage suited to their conditions, it is not essential that the stream should be continuously, at all seasons of the years, in a state suited to such floatage".* If small boats could have been used to transport people and local products or supplies, this capability supports a finding of navigability.

The courts ruled that capacity for navigation, not usage for that purpose, determines the navigable character of waters with reference to the ownership and uses of the land covered by the water. The products of the community at least in some considerable measure may be transported upon the waters if so desired, and the waters are admittedly of considerable area and useful for general navigation in small boats containing persons engaged in pursuits either of business or pleasure. The fact that a lake goes dry is unimportant if in its ordinary state it is in fact navigable.

Ordinary High Water Line (OHWL)—On Florida's nontidal navigable lakes and rivers, the boundary between sovereignty submerged lands and private uplands is the OHWL. Most modern definitions of OHWL have been derived from the concurrence in Howard vs. Ingersoll, a United States Supreme Court case in which Justice Curtis wrote: *"The line is to be found by examining the bed and banks, and ascertaining where the presence and action of water are so common and usual, and so long continued in all ordinary years, as to mark upon the soil of the bed a character distinct from that of the banks, in respect to vegetation, as well as in respect to the nature of the soil itself. Whether this line... is found above or below, or at a middle stage of water, must depend upon the character of the stream".*

The high water stages of rivers and lakes are unpredictable and variable. Thus, in nontidal waters, the emphasis is on drawing a line that reflects the physical effects of high water stages on the land. This line is determined on a case-by-case basis. In situations where the banks are low and flat, the water does not impress on the soil any well defined marks of demarcation between the bed and the banks. In such cases the effect of the water upon vegetation must be the principal test of determining the location of the high-water mark as a line between the riparian owner and the public. It is the point where the presence or action of the water is so continuous as to destroy the value of the land for agricultural purposes by preventing the growth of vegetation. Areas that are subject to flooding of such frequency and duration as to prevent the growth of ordinary agricultural crops are clearly below the OHWL.

Vegetation—A number of court cases have addressed the relationship between the OHWL and vegetative indicators. In a leading case, the federal government's construction of a dam raised water levels on a river, damaging a city's gravity-flow sewerage system. The question of damages required the court to determine the OHWL before the dam was closed. After approving the general definition of the OHWL, the court stated: *"The vegetation test is useful where there is no clear, natural line impressed on the bank. If there is a clear line, as shown by erosion, and other easily recognized characteristics such as shelving, change in the character of the soil, destruction of terrestrial vegetation, and litter, it determines the line of ordinary high-water...Also a test of the distinct line is the destruction of terrestrial vegetation so these are not really two separate tests but must, of necessity, complement each other".*

The courts have not developed standards for determining species dominance in reference to the vegetation test. Because several methods exist for defining "dominance"—including measures of basal area, relative numerical density, and relative cover—any vegetation analysis should include data and discussion of each possible aspect of the definition. Some courts too readily accept an expert witness's opinion although a sufficient explanation of the method used to arrive at the stated conclusion is not provided.

A federal case, United States vs. Cameron, from Florida, illustrated several points regarding the use of vegetation surveys in boundary determinations. A rancher owning land along a lake connected to the St. Johns River constructed an extensive dike on the low-lying portions of his property. He also installed a pump to remove water from the diked area. The Corps of Engineers charged him with violating the Rivers and Harbors Act, claiming he had constructed the dike below the OHWL without a permit. As part of the evidence presented, both parties submitted analyses of the vegetation above the dike and on the exposed land between the dike and the lake.

The government's expert witness limited his investigation to submerged vegetation. He found spikebrush was the dominant species in a significant portion of the area both inside and outside the dike. He testified that spikebrush, as well as four other species he identified, live in areas that are inundated over fifty percent of the time. At the site, elevations below two feet were normally inundated fifty to seventy-five percent of the time, and the topography of the area showed that more than one-third of the site was below the two-foot elevation.

The rancher's expert witness examined the site slightly over one year after the government's witness completed his survey, and one year after a pump was installed to drain the land. The expert's analysis focused primarily on the existence of upland vegetation, and he found predominantly upland species on both sides of the dike.

According to this witness, the dominant species were Bermuda grass and Bahia grass, both "transition zone" species that could survive total inundation for no more than two to twelve weeks. He also found that cypress trees on the lakeward side of the dike had not formed new knees, suggesting that these areas had not been inundated for significant lengths of time in recent years. The rancher's expert ultimately concluded that the area on either side of the dike had not been submerged a significant amount of time in recent years.

The court recognized discrepancies in the evidence. After weighing the evidence, the court concluded that the dike was in a dynamic marshland environment which was experiencing a period of low water. The court attributed the discrepancies in the vegetative analysis to the natural changes in water level that occurred in the year between the two surveys. The court also noted that the rancher's expert witness performed a soil analysis which indicated greater amounts of organic matter in the soil as one moved from the lake to the dike. However, the court seemed to focus on the existence of upland vegetation lakeward of the dike, holding that this evidence precluded any possibility that *"such area is so usually covered by water that it is wrested from vegetation and its value for agricultural purposes destroyed"*.

This case illustrates the difficulty of reconstructing normal conditions in a disturbed area. Neither the court nor the parties involved acknowledged that construction activities, pumping, and the effects of the dike itself could have drastically altered the water regime in the area. Vegetation and soil analysis conducted at several nearby sites would have allowed a reasonable, and probably more accurate, extrapolation of the OHWL. (Writer's opinion: this case was decided years before the Sabal Hammocks case—however, it strikes of unique similarities in the 2005 ruling on Sabal Hammocks.)

The court's willingness to accept evidence of "some" upland vegetation as indicating agricultural usefulness below the level of the dike is significant. Neither party made any attempt to identify the relative densities of upland and submerged species. In this regard, the government's expert witness could have provided more documentary support for his vegetation analysis and improved his credibility if he had included surveys of both submerged and upland species and the degree of dominance of each along several transects in the area.

Conclusions—Florida's Governor and Cabinet, who sit as the Trustees of the Internal Improvement Trust Fund have the authority to settle boundary disputes. The Florida legislature should consider legislation authorizing the adoption of rules that incorporate a more definitive procedure for establishing OHWLs. The Trustees already may have such authority. A Trustees rule at least could establish the conditions under which the Trustees will agree not to contest an OHWL survey. Such a rule would go far toward giving landowners the certainty they desire. The rule that the Department of Natural Resources proposed in 1988 was generally consistent with the existing case law and should be resurrected as a starting point for the adoption of any rule.

Other proposals that have been made, such as the rule adopted by the Board of Professional Land Surveyors and the one proposed by the Governor's Ordinary High Water Line Committee would constrict severely the boundaries of the public trust. Landowners' advocates will probably attempt to use any efforts at legislative reform or rulemaking to adopt the substance of the latter proposals. Extraordinary vigilance is necessary to ensure that efforts to clarify the methodology are **not used** as a subterfuge to divest the people of their public trust and natural heritage in submerged lands. (This concludes the data provided from "Florida Law Review".)

In the following two sections of this chapter, I will discuss my personal involvement in requesting an OHWL survey of the Duda Ranch, via the Brevard County Commission. The ranch is located along the St. Johns River in Brevard County. Also, I will discuss my position in obtaining an OHWL survey of the proposed Sabal Hammocks project in Brevard County on Lake Poinsett.

The Duda Ranch Episode

This writer initiated action to obtain an OHWL survey of the 38,000-acre Duda Ranch, which was in the early stages of being transformed into the new city of Viera, located north of Melbourne along the I-95 corridor, on the east and west side of the Interstate in Brevard County. On the western side of I-95, the ranch property stretched over 5 miles to the eastern shoreline of the St. Johns River. I worked with the Division of State Lands, Brevard County Commission, the SJRWMD and the Florida Fish and Wildlife Conservation Commission, exchanging letters with these agencies and providing boat tours of the 14-mile stretch of the Duda Ranch along the St. Johns River.

Moccasin Island, Lake Winder: part of 14,137-acre state purchase of Duda Ranch—check out the Sand Hill cranes; Joyce and I counted over 100

In a letter to this writer dated 13 April 1994 from Henry Dean, Executive Director, SJRWMD, Mr. Dean stated that while it had not been the policy of the SJRWMD to intervene between a private interest and the state, the SJRWMD would provide technical support to the state if the state decided to perform an OHWL survey of the ranch.

On 1 November 1994, I addressed the Brevard County Commission regarding an OHWL survey of the Duda Ranch. I informed the board that I had been authorized to speak on behalf of the Florida Chapter—Sierra club, the Florida Wildlife Federation, Indian River Audubon Society, the Spacecoast League of Women Voters, and SAVEs 25 member organizations. I reminded the commission that in March 1993, at my request, the board notified the state by letter that Brevard County Government was opposed to private development on state sovereign lands.

At a subsequent meeting, following a discussion in which Attorney Mason Blake, representing Viera's interest (Duda Ranch) was present, the board voted 5-0 to forward a request to the state seeking an OHWL determination of the diked areas of the ranch which encompassed thousands of acres of the property. The commission's action put future expansion and planning for the city of Viera in a state of uncertainty.

Three months later, on 1 February 1995, I addressed a "work shop" session of the Brevard County Commission. My purpose of speaking was to ensure that the commission was engaged with the state in securing information in support of an OHWL survey of the Duda Ranch. As I was speaking, Commissioner Nancy Higgs inquired if I had seen the Viera Company letter dispatched to the Secretary of the Department of Environmental Protection (DEP) dated 31Janaury 1995. When I replied that I had no knowledge of the letter, Commissioner Higgs requested the Viera Company provide me with a copy for my review. I was promised I could continue my presentation following my review of the several attachments to the letter to the DEP Secretary.

Following my review of the letter with its attachments, I again addressed the commission. I stated that the Viera letter signed by Viera Company Vice President, Corporate Council R. Mason Blake was, in my opinion, an attempt to stop any action to proceed with an OHWL survey of the Duda Ranch. The Viera letter indicated that such a survey had already been conducted in 1969. The 1969 survey was a Mean Water Line survey which depicted normal water levels, and not "ordinary high" water levels.

The following day, 2 February 1995, I dispatched a letter to the Secretary, DEP. I informed the Secretary that the Viera Company's letter to her was discussed the previous day before the Brevard County Commission. I further expressed that the county commissioners wanted to proceed in negotiating an OHWL survey of the Duda Ranch. I provided the commission a copy of my letter to the Secretary. At this point, Viera officials understood the county commission would proceed in requesting an OHWL survey at a point in time before permitting any further expansion west of the initial 5,800-acre construction site.

During a subsequent appearance before the commission, I offered a typewritten paragraph I had prepared prior to the meeting, requesting the same be included in the Viera Development Order. The paragraph stated that no further expansion westward would be permitted until an OHWL survey was performed on the low-lying areas of the ranch, along the St. Johns River. Following a review by county staff personnel, the commission voted to insert proper language into the development order.

In addition to this writer's recommendation, the commission's language included a 322-foot setback line landward of the state's location of the OHWL. The Viera Company was officially put on notice; suspected areas of sovereign lands would be surveyed prior to further development westward toward the St. Johns River.

In 1999, with the need to plan whatever future development would be permitted, rather than agree to an Ordinary High Water Survey, the Viera Company sold 14,137 acres of the Duda Ranch to the state for $24.8 million dollars. Cost per acre was approximately $1,700. In this writer's opinion, Viera officials determined their best option was to sell the questionable lands that would be the focal point of any state OHWL survey for potential sovereign lands.

Prior to the official state purchase, the SJRWMD provided this writer with a seven page proposal regarding the purchase. I was requested to make any comments I felt should be considered in the purchase agreement. The first two words of my comments were: "Buy it". I requested fence lines be removed, along with the "No Trespassing" signs, and temporary poles be placed along the property line depicting the 100-year floodplain boundary. The state's planned purchase included all lands up to the 100-year floodplain.

Subsequently, several SAVE officials were provided a tour of the soon to be "public lands". The lands included 14 miles of shoreline up to the 100-year floodplain of the St. Johns River. Indeed, it was a significant victory for the "people". A very special thanks to Commissioners Nancy Higgs, Truman Scarborough and Karen Adreas for support of SAVEs position to protect potential sovereign lands for future generations to enjoy.

During SAVEs tour of the property, I discovered a unique relic from the past. The "facility" was almost totally overgrown with vegetation. At first look, I thought it was might have been a small barn that had been almost destroyed by inclement weather over the years. Upon taking a closer look, I realized exactly what once had occupied the space, well landward of the present day dikes.

This writer requested the Duda Ranch tour guide to explain the almost invisible "shack". He stated he had no idea, but whatever it "was", it had been built there many, many years ago. I pointed out to the SAVE officials how I determined what once was perhaps a highly used building. I saw what had been a large opening facing west toward the river. As I removed vegetation along a path directly in front of the opening, I discovered a number of rotted pilings, still visible and partially in place. This area once offered a boat dock and a boathouse—explaining the reason for the open area facing the river. I estimated it was abandoned 55 to 60 years earlier, about the time the startup of dike construction at the former Duda Ranch.

Proposed Sabal Hammocks Project OHWL Survey

Chapter V addressed how this once-proposed project evolved and why SAVE filed legal action against the proposed project. The following paragraphs present the issue of private lands vs. sovereign (public) lands at the proposed project site. In 1990, SAVE hired Consultant Dennis Auth and Water Quality Expert Dr. Woody Deirberg to investigate potential sovereign lands, including soil samples on the 720-acre site located behind an 18,000-foot illegal dike on the St. Johns River, in Brevard County. Initial findings determined that a minimum of 160 acres of the project was planned for construction on state-owned submerged lands of Lake Poinsett. These submerged lands were kept dry most of the year by the use of extensive pumps mounted atop the illegal dike.

During SAVEs legal challenge before the Governor and Cabinet in November 1991, SAVE Attorney Tom Tomasello, with the assistance of Attorney General Bob Butterworth, obtained the support of the Governor and other Cabinet members to dispatch Division of State Lands surveyors to the site for a "safe line" determination. The state preliminary findings indicated that a minimum of 243 acres of the project would be found to be state-owned sovereign lands should an OHWL survey be performed. The owner/developer was instructed not to proceed with any development activity on the property until the sovereign lands issue was settled.

In 1995, the state took over the legal challenge SAVE had pursued for five years. The actual OHWL survey revealed there was a minimum of 272 acres of sovereign submerged lands in the proposed project site. The developer had been granted several extensions to allow him to prove the lands were indeed above the OHWL.

In 1977, the developer's own surveyor placed the OHWL line at 13.0 ft el. Finally, in January 2005, the case was heard before Judge Bruce Jacobus, Eighteenth Judicial Circuit Court in Brevard County. Joyce and I sat through the 5-day trial, feeling confident Judge Jacobus would find the state's agreement to accept the 13.0 ft el OHWL survey of 1977 to be fair. In fact, the state OHWL survey positioned the OHWL at 13.8 ft el. The illegal dike had been constructed at the 12.1 ft el in 1973. Therefore, the 272-acre claim of sovereignty by the state was a reasonable finding of fact.

In the Final Order dated 25 August 2005, Judge Bruce Jacobus cited the legal definition of OHWL as detailed in Tiden v. Smith (Fla 1927) and adopted by the Fifth District Court of Appeal in the Board of Trustees of the Internal Improvement Trust Fund v. Walker Ranch General Partnership (Fla. Fifth DCA 1986). Following are the highlights of the Final Order as written by Judge Jacobus:

"Definition, Page 4, Final Order

In some places, however, where the banks are low and flat, the water does not impress on the soil any well-defined line of demarcation between the bed and the banks. In such cases the effect of the water upon vegetation must be the principal test in determining the location of high-water mark as a line between the riparian owner and the public. It is the point up to which the presence and action of water is so continuous as to destroy the value of the land for agricultural purposes by preventing the growth of vegetation, constituting what may be termed an ordinary agricultural crop.

Burden of Proof, Page 5, Final Order

An ejectment action seeks to oust from the property the person in possession. The Trustees...have the burden of proving by a preponderance of the evidence title and right of possession to the property involved. To prevail on their claim, the Trustees must prove the location of the ordinary high water mark...of Lake Poinsett is thirteen feet.

Stage Duration, Page 5, Final Order

Stage duration is a statistical analysis of the water levels of a water body to determine how long it remains at a certain elevation on an annual basis. As it relates to Lake Poinsett, statistical data exists...some land experts have proposed that by performing a statistical calculation using the collection elevation data, the State should adopt, as the ordinary high water mark, the elevation of the lake at a certain percentage of the time...This is particularly true on lakes such as Lake Poinsett where the elevation change from the edge of the lake landward is minute over a long distance. That is, the property is relatively flat, so a small change in elevation moves the edge of the lake a substantial distance from its normal boundaries.

This Court did not permit testimony as to stage duration; however, the Court did allow the statistical data into evidence during the trial. The exclusion of the stage duration testimony was...there is no acceptance in the scientific community that the 30% stage duration equates to the ordinary high water mark. This was substantiated by testimony of George Cole, who is a professional land surveyor and formerly the State's cadastral surveyor...Mr. Cole was a proponent of the stage duration method...however, his testimony at trial was, and the Court accepts it as credible, that he has receded from the view as further studies have revealed...there is absolutely no correlation between the 30% stage duration and the ordinary high water mark.

Ordinary High Water Mark, Page 7, Final Order

...quite a bit of information exists regarding Lake Poinsett and its boundaries, from the late 1800's to the present. The difficulty in this case in determining the ordinary high water mark is that the land is relatively flat. There is very little elevation change from the edge of Lake Poinsett in any direction for long distances. Terry Wilkinson, who is currently the cadastral surveyor for the State of Florida, testified that the ordinary high water mark is at the thirteen (13) foot elevation. It is interesting that this corresponds to the 30% stage duration, and is the upland boundary of the wetland. On the other hand, Mr. Smith, through his surveyor, Daniel Gentry, alleges that the ordinary high water mark is approximately 12.1, and he contends that is where it has been through the history of the lake.

Another complicating factor is that the Smith property, which is the subject of the litigation, has three dikes, built in 1954, 1969, and 1993 (Writer's Note: 1993 should have read 1973). *Some of the dikes were breached later in time; however, the dikes cause the impoundment of water on the upland side from run-off on the property.*

The Court heard testimony from numerous experts in the field of soils, plants, vegetation, and the type of crops that could be grown or not grown on the property...the Court does find compelling, the work of Daniel Gentry, as it relates to the Smith property. Mr. Gentry ran some transients on the property which show that there is a mark on the bank of the lake...the weeds go from those that are aquatic, such as bulrush and spartina, to those which are not, such as, willows. That is true for the transects he ran on the Smith property, and those transects show the mark on the bank is at the 12.1 elevation...the Court finds that they support the boundary of the lake at the 12.1 elevation.

Accepting the Trustees' position that the ordinary high water mark is at the thirteen feet NGVD, might have some support as it relates to the Smith property, in that the thirteen feet contour places the edge of the lake in some places, almost three miles from the edge of the lake that a person of common intelligence would clearly recognize as the boundary of the lake. For example, utilizing the thirteen feet contour would almost completely encompass the subdivision that was the subject of Mr. Cole's letter in 1980, resulting in that property being constructed on the sovereign land of the State of Florida.

In reviewing the lake in its totality, the thirteen feet contour does not comport with logic and reason. Therefore, the Court finds that the Trustees' position that the ordinary high water mark of Lake Poinsett is at thirteen feet NGVD is not supported by the greater weight of the evidence...therefore the Court finds that the ordinary high water mark is at 12.1 feet NGVD; the Court accepts the boundary of the Smith property between the sovereign lands of the State of Florida and the private ownership of Smith, is as set forth in the survey prepared by Daniel Gentry, dated March 3, 2001". (End of excerpts from Final Order)

In the opinion of this writer, justice failed the citizens of Florida, especially those living in Brevard and the surrounding counties who frequently use the waters of Lake Poinsett and the St. Johns River. Before electric trolling motors were invented, one either paddled their boat, drifted aimlessly with the wind, or anchored their boat and waited for the fish to show up. During the late 1950s and 1960s, this writer, joined by other anglers anchored our boats near the area of the illegal 1973 dike, got out and waded the flats of Lake Poinsett. The very area of this litigation provided this writer and others, many enjoyable outings and successful fishing trips well landward of the 1973 dike. Before this dike was constructed, the litigated area was also used for many years by duck hunters.

This writer remains convinced the state would have prevailed in its case if the decision had been made to continue the case before the State Supreme Court. Furthermore, I am convinced that a person of "common intelligence"—assuming they had a basic understanding of the state constitution, would agree that in many places the "flats of Florida" extend much more than 3 miles, such as the Everglades. On a positive note, I am encouraged at the potential state or county purchase of the submerged lands on the property.

This writer disagrees with Mr. Gentry's finding of a "bank" along the east shore of Lake Poinsett. It is the opinion of this writer that Mr. Gentry mistakenly "found" the relics of the failed 1969 meandering dike that was abandoned because action by the waters of Lake Poinsett destroyed the dike, leaving a somewhat "elevated or berm area" along the northeast shoreline of the lake. The 1969 dike was an earlier attempt to remove the public from it constitutional right to enjoy the use of public lands. The final order issued by Judge Jacobus did not set any precedent, since the case did not proceed to the State Supreme Court.

Over 70 percent of the 720-acre site is located in the annual to 10-year floodplain, separated only by the illegal dike which, in this writer's opinion, would not survive a significant hurricane. Brevard County ordinances do not permit development in such areas. Mr. Smith, the owner of the property, has indicated he is willing to sell the land to the state. I understand there is action underway to review the potential purchase for conservation, including removal of the illegal dike.

Proposed Legislation threatens Public Ownership of Navigable Lakes and River

Recent attempts by some state legislators called for changing the definition of the OHWL on Florida's lakes and rivers. These events occurred during the 2000 and 2005 legislative sessions. The sponsors of the legislation stated their intent was to "clarify" the location of the OHWL. Had these attempts for "clarification" become law, the Florida Statues would have been amended to specify the OHWL would be the mark on the land where the waters of the rivers and lakes were present "a majority of the year". That is in direct conflict with present law which places the OHWL at a point where the water "reaches during the wet season".

Typically, the water would be present only during the several months of summer and not the minimum of over 6 months as required by the proposed legislation. In many of our lakes and rivers in Florida, such action would have eliminated hundreds of thousands of acres of "public" lands.

This writer was only one of hundreds who received an ALERT published by David Guest, Council for Earth Justice Legal Defense Fund in early February 2000. Mr. Guest was a former Florida Assistant Attorney General. I quote from Mr. Guest's ALERT notice. *"The Florida Farm Bureau and large timber interests are pressing for new legislation that would convert hundreds of navigable lakes and rivers into private property.*

Under long-standing law, enshrined in the Florida Constitution, the navigable waters are held in trust for public use such as boating, fishing and hunting. If the proposed legislation passes, timber companies would be able to cut down the cypress forests that line many of the state's waterways, while agricultural interests and developers would be able to drain or fill shallow marshes that often line the shores of waterways.

....The Farm Bureau and timber interests will argue that they are unfairly paying property taxes on navigable lakes and rivers without enjoying the full privileges of ownership...Soon after bulk deeds were issued, the Florida Supreme Court was presented with the question of whether swamp deeds conveyed navigable waters. The Florida Supreme Court said resoundingly "No" almost a hundred years ago". That ruling continues to be the law and the position of the Florida Supreme Court.

On Saturday, 26 February 2000, I contacted SAVEs 25 membership organizations. Subsequently, the State Capital was bombarded with phone calls, fax message and letters in opposition to the "now identified" bills. Other conservation/environmental groups were taking similar action. In the State House of Representatives, Rep. Paula Dockery sponsored HB 1807, while in the State Senate, Senator Walter Campbell sponsored SB 1824.

These bills, referred to as "Florida Land Title Protection Act" would in fact, be a land swindle by cattle, timber, and agricultural landowners who were attempting to gain ownership of sovereign lands held in trust for the "public".

Governor Bush joined Attorney General Butterworth, Comptroller Milligan, Insurance Commissioner Nelson and Education Secretary Gallagher in opposing such an embarrassing public land-grab act. The bottom line contained in these bills would have resulted in the relocation of the Ordinary High Water Line to the present day location of the ordinary low water line. In Florida's flat topography, the results would have amounted to the loss some 500,000 acres of present day public use lands. Over19 major conservation/environmental groups put in hundreds of hours of hard work to prevent the proposed bills from becoming law.

SAVE received positive responses from the Governor and Cabinet officials, who listened to the overwhelming voice of our supporters. These great servants to the people also understood the 19 major conservation and environmental groups across the state, representing tens of thousands of citizens who registered their opposition to HB 1807 and SB 1824.

This writer's personal files are filled with many letters SAVE exchanged with conservation and environmental groups from around the state in our determination to stop the "taking of public lands". I am convinced that on the final day of the 2000 legislative session, Senate President Toni Jennings understood Attorney General Butterworth's words: *"If these bills are enacted, I could not defend them in a court of law; they violate the State Constitution"*. Senator Jennings was also remindful of the tremendous outpouring of opposition to such takings of public land. To her credit, Senator Jennings would not permit Senate Bill 1824 to be heard. The session was over.

The legislative session was over. Protection of existing state sovereign land remains in place for the use of all citizens. The friends of our lakes and rivers celebrated this hard fought victory. It was a team effort from the outset. I was privileged to be a part of the dedicated group of "volunteers" from around the state.

This writer extracted a part of my letter printed in the Guest Column of Florida Today on 17 July 2000, which reads as follows: *"Florida's Public Lands Held in Trust for All—With their land-grab agenda well planned, the cattlemen's association, timber, the farm bureau and some developers could not overcome the public voice that was heard in Tallahassee during the past legislative session. Without the determined efforts of Attorney General Bob Butterworth, Senate President Toni Jennings, and Senate Majority Whip Jack Lavatla, the revision setting the ordinary high-water mark legislation could have become law"*. (Writer's note: The voice of the people was significant in defeating these bad bills.)

The Florida House of Representatives highly controversial bill (HB 1807) had been passed by a vote of 70-45. I thank the 45 members who voted against the taking of public lands.

Brevard's legislative delegation on the House side yielded to the lobbying effort of special interest groups. State Reps. Randy Ball, Howard Futch, Harry Goode and Bill Posey, along with Rep. Charles Sembler of Indian River County, all voted in support of this bad legislation.

No deed or title conveyed by the state entitles the upland landowner to deny public access to any water body below the ordinary high water line. Also, it's worth noting that Rep. Tom Feeney of Seminole County, the next House speaker, voted to give away the public lands.

Senate Bill 1804 had been co-sponsored by Senator Charles Bronson. In fact, following the action by Senate President Toni Jennings, Senator Bronson stated those special interest groups should file a class-action lawsuit against the state. Better heads prevailed; no such lawsuit was filed.

2005—Another Legislative Attempt to Transfer Public Lands into Private Ownership

After a 5-year absence, once again legislation was filed to redefine the Ordinary High Water Line. The code words were labeled: "Clarification of the Ordinary High Water Line". This approach was really no different, in purpose, than the 2000 legislative effort described in the preceding paragraphs. Again, the conservation/environmental community was strongly committed to defending the public's right to access public trust lands under present law.

Again, SAVE was among the many organizations that informed the sponsors of the ill-advised legislation, that we opposed their bills. We would challenge any attempt to change the present law.

In the following paragraphs, I could repeat the year 2000 saga and be on target with this new legislation. I obtained copies of the House and Senate bills. The bills sponsor names changed, but the intent was similar to the failed attempt in the 2000 legislative session.

Senator Bill Posey, in a letter to this writer, dated 8 November 2004 stated how much he appreciated my letters to the Editor. In his words, he stated *"only the best and most meaningful letters out of all the thousands written are selected to be published. Thank you again for making the effort and helping me to remain informed about the issues that are important to you....Being an effective legislator depends upon the ability to stay informed about the issues that are of concern to the citizens of Florida".* This writer sincerely appreciated his kind words, but once again the citizens were being ignored in order to satisfy the special interest groups; cattle, timber, agriculture and developers.

Senator Posey sponsored Senate Bill 1954; it was entered into the record on 2 March 2005. The companion bill in the House of Representatives was HB 1369, sponsored by Representative Seiler. After reading SB 1954, I contacted Senator Posey by letter, dated 11 March 2005. I quote, in part, from my letter to the Senator:

"I want to address the peoples' concerns about SB1954:....1) loss of public access to the marsh of the St. Johns River in your district;....2) loss of wildlife habitat, especially spawning grounds for freshwater bass, which typically spawn in one to two feet of water...loss of feeding areas for wading birds, and the endangered snail kite;...3) building to the water's edge of large scale condos and houses with their damaging impact to the water quality of the St. Johns River;...4) violation of the Florida Constitution, Section 10, Article 11, which protects the publics' right to utilize navigable waters of the state;...5) loss of large portions of marsh areas which is filled during the wet season and remains so for months providing filtering of runoff from storm events and agricultural operation.

There are many other bad effects that would result should these bills be enacted by the legislature...listen to the people. I am available to speak with you or your aide on this matter".

This writer never received a response from Senator Posey.

My letter to Senator Posey was faxed to all members of the Brevard County Legislative Delegation, Governor Jeb Bush, Attorney General Charlie Crist, Solicitor General Chris Kise, Florida Audubon Society—Charles Lee, Earth Justice Legal Defense Fund—David Guest, Brevard County Commissioner—Sue Carlson, and the Florida Wildlife Federation—Manley Fuller.

Senate Bill SB 1954 language in part reads: "Section 253-024, Florida Statutes, is created to read:

Ordinary high-water line.—This section applies when constructing the term "ordinary high-water line" as it is used in this chapter. The term "ordinary high-water line" includes the terms "ordinary high-water mark." "line of ordinary high water," and "ordinary high watermark." The ordinary high-water line is the visible

mark formed on the bank of a fresh waterbody by the continuous presence and action of water where it

stands for most of the year and wrests the bed of vegetation."

The wording in Senator Posey's bill would render hundreds of thousands of submerged lands in Florida into ownership of upland landowners. The key words "action of water where it stands for most of the year" is significantly different from the law that presently exists—"action of water where it reaches during the wet season". **Senator Posey's bill failed to clear committee hearings.**

In an article published in Florida Today on 23 March 2005, Senator Posey stated his bill would "clarify" the high water line. In the same article, I stated that *"if you say 'where the water level is most of the year, then you're down totally out of the marsh grass, you're down in the lake bed".* Senator Posey also stated he is simply trying to clarify how the state defends the high-water line, which he says has been subject to differing interpretations by the courts during the years. On this matter, I agree with Senator Posey; however when it comes to interpreting the State Constitution, my position swings to the court, who should be more qualified and have no pressure from lobbyists as to what needs to happen in the court room.

Such is not the case in the halls and chambers of state government proceedings. Lobbyists play crucial roles in their appearance before state officials in Tallahassee; often successful in influencing the decisions rendered by government.

On 30 March 2005, Florida Today Editorial Page Editor John Glisch reiterated the environmental community's position. The headline read: "STOP THE LAND GRAB". In the subheading, the article began with the words: *Senator Posey's bill would rob Florida citizens, reward developers and industry".* My personal thanks to Mr. Glisch for his strong editorial. My own, but similar statements were published in the Florida Today Guest Column on 6 April 2005. This writer has a collection from newspapers around the state, all of which agree with the people—do not attempt to remove the people from these public lands.

Future legislative sessions should deal with the critical issues facing all Florida citizens: healthcare, education, crime, the environment, hurricane impacts, tourism, and in general—listening to the voice of the people.

On 12 May 2005, I wrote a letter to SAVE membership organizations explaining how the organized efforts of groups like SAVE had prevailed over the far less number of big business lobbyists with deep pockets. The initial "fire" ignited at the start of the legislative session was soon extinguished by the public outcry for the protection of our rights since statehood to use the sovereign public lands of Florida. Attorney General Crist informed the legislators that the defeated bills could not have been defended in a court of law. Enough legislators agreed with the Attorney General and the public outcry to kill the ill-advised bills.

Once a nice private camp: removed due to liability issue; was built on potential sovereign land

In summarizing the complex issues of determining ordinary high water lines on the rivers and lakes of Florida, I submit that such determinations are a complex issue. However, any citizen has the constitutional right to express an opinion on such matters of law. I encourage the reader to become knowledgeable of issues that determine the future environmental protection of our rivers and lakes. Public libraries are a great source for information; the Florida Statues are readily available.

More Fishing Trip Memories—On a hot summer morning in July 1989, I guided a father and son team from New York City for a half-day trip on Lake Poinsett. Dad assured this guide that he and his son were familiar with bass fishing. I noticed their expensive fishing rods. We would be fishing with the popular artificial plastic worms. In less than 10 minutes from the boat ramp, we were at my pre-selected spot, an open area near an island on the southeast shoreline. (One of my fishing pals, David Solomon had informed me that he had named that area *"Leroy's Island—every time I'm own the water, I see you fishing that island".)*

This was going to be a great day of fishing. Within the first 30 minutes, my guests had caught 5 nice bass. The son said to me: *"Why aren't you fishing?* I responded that I was here to see that they had a good time, I could fish another day. He and Dad asked me to join them. They didn't have to repeat the offer. I had a pre-rigged rod and reel ready. In short order, we all three were catching fish. At one point, I laid my rod down to clean my sunglasses.

Unknowingly, my plastic worm was dangling just under the surface of the water. I heard a noise—it was my rod about to slide off the boat into the water. I quickly grabbed it; there was a bass on the other end. I landed this gift; it weighed 4 pounds. At the request of my guests, we placed about 10 of the larger bass in the live-well for photos at the end of our trip. Following the photo-opt, all bass were released back into the St. Johns. Following lunch at a nearby restaurant, my guests were off for their return trip to New York City.

CHAPTER VIII—

TRANSFORMING DUDA RANCH—CITY OF VIERA

In Brevard County, a new city is born. Everywhere you look, the amenities abound. Numerous beautiful waterfront homes are only a stone's throw from thriving new businesses, including shops and restaurants. New schools, churches, and a VA Clinic are up and running. The Brevard County Government Center was among the first buildings constructed.

Recreational opportunities are numerous. One golf course is in operation, and another one is planned. Lighted outdoor playing fields exist for a variety of sports activities as well as a major league baseball "spring training" facility. The nearby historic St. Johns River provides nature lovers with boating, fishing and "gator viewing". This river is a wild and beautiful paradise, and a part of my very soul.

SAVE was very concerned with the building of a new city on former ranch lands. The St. Johns River bordered 14 miles of the ranch on the east side of the river. Negative environmental impacts to the river were foremost on the mind of this writer. These concerns would play out during the early planning phase for the new city. The East Coast Regional Planning Council granted SAVE "reviewer status" during the pre-development process; commonly known as "Development of Regional Impact" (DRI). SAVE authorized this writer to represent our interests, document and question any issues of concern.

As I reviewed pre-development vs. post-development requirements stated in the DRI, I was convinced that post-development conditions would significantly impact the water quality of the St. Johns River. Lake Winder would be the primary receiving waters of the direct discharges from the development. My concerns were supported by historical data from the Florida Fish & Wildlife Commission (FWC). Specifically, property below the 15-foot contour line of the Duda Ranch included several thousand acres I believed to be sovereign (public) lands. In addition, I suspected part of the mining area where road material would be excavated to be potentially sovereign lands.

Topography maps of the ranch provided by FWC convinced this writer that development in any area where the contour line was 15 ft. el, or below, was suspect. An FWC map, dated in the 1920s, clearly identified numerous lily pad fields on the ranch up to the 15-ft contour line. This map was presented by FWC for review and discussion with the Brevard County Commission during the DRI review process.

I recall a visit Joyce and I made to the real-estate office of the Viera Company. This writer wore a suit, with coat and tie. Since my retirement from the Aerospace industry in 1989, I wear a suit only to funerals and weddings. I made an exception for this outing. When the Viera "investment" representative inquired of our particular interest, I indicated our interest was in "a potential future investment". My response regarding a future investment was in fact, true. However, that "future investment" involved the protection of the St. Johns River, not in development.

Joyce and I were escorted to a room where a display covered almost an entire wall. There we observed the planned total build-out of the soon to be "former ranch". That plan revealed a park to be constructed at Moccasin Island, located on the water's edge at Lake Winder. We were informed the park, when completed, would be donated to Brevard County. Also, waterfront property development was a long range plan, at least 15 years out. I stated to the Viera representative that such a time period was beyond my interest, and we quietly left the office. I had accomplished my purpose for the visit. Indeed, future plans would locate development below the 15-foot contour line.

At this point, I decided to "find a way" to protect potential sovereign lands and the annual to 25 year floodplain of the St. Johns River bordering the planned development.

The reader will discover how I gained support of the Brevard County Commission to place certain language in the Viera Development Order, which played a key role in the state's subsequent purchase of 14,000-plus acres of the Duda Ranch along the St. Johns River. For years, the ranch owners had been unwilling to sell any part of their 38,000-acre property.

The new city of Viera is a well-planned development. The "town center" is the heart of the city. Upon completing the build-out, over 40,000 residents will occupy the former ranch. Many of the city's commercial facilities and recreation fields can be seen as one travels Interstate 95 south through Brevard County.

Joyce and I frequently visit the area. We are in awe of the newly created infrastructure. On our visits, we travel on new roads, observe newly created lakes, churches, schools, etc. How many people have actually watched the construction of a totally new city? More important to this writer, I am convinced the city will have minimal environmental impact when compared to the pumped agricultural discharges of polluted runoff from the former ranch. The state's purchase of former ranch lands included all lands up to the 100-year floodplain.

As a result of the state's 14,000-plus acres purchase, the environmentally sensitive lands on the Duda Ranch are now in public ownership. Proper conservation measures are in place to protect this 14-mile stretch of the natural beauty on the St. Johns River, while affording the landowner the opportunity to realize the success of a well-planned community. The following pages offer the reader a look at a part of the past history of the Duda Ranch and this writer's determined effort to ensure the new city of Viera would negatively impact the St. Johns River.

Illegal Modifications to Canal System on Duda Ranch

In the early 1990s, this writer pursued major violations of established permitting requirements at the Duda Ranch. Illegal modifications to the massive drainage system of canals affected over 25,000 acres of the Duda Ranch. These violations occurred over a three-year period; 1989, 1990 and 1991 and adversely impacted the water quality of the St. Johns River. The SJRWMD was unaware that major culverts were added to expedite the movement of stormwater from the Duda Ranch to the St. Johns River. In addition, canals were widened to accommodate the increased flow. The following chronology describes the incredible events which occurred over a three-year period.

In 1987, the Department of Environmental Regulations (DER) issued consent orders to all farming and ranching operations along the Upper Basin of the St. Johns River. Agricultural interests were given 5 years to construct on-site wet-retention reservoir systems to collect farm runoff. These large "holding reservoirs" would be used to store farm runoff. The water could be reused for irrigation purposes. All farms, except one, complied with the order. Their reservoirs were in place by 1992. The one exception was the Duda Ranch.

The SJRWMD pressed the ranch into action. Finally, in 1994, ranch officials submitted an application for the construction of a 452-acre reservoir to be located inside their diked property. Attached to the application (at the direction of the SJRWMD) was an "after the fact" permit applica-

Cattle grazing in areas of the Upper Basin marsh; they are a major agricultural investment

tion for the 3-year period of unauthorized drainage of the controversial 25,000 acres.

Certified letters from the SJRWMD to A. Duda & Sons, Inc were dispatched. My earliest copies of these letters date back to 1992. In one letter dated 12 March 1992, the SJRWMD informed ranch officials that *"we have reviewed the available water quality data and...determined that discharges from the ranch cause or contribute to violations of state water quality standards....District staff has determined that the pumps contribute a significant portion—greater than 5% of the phosphorus load to the river"*.

On 4 November 1992, BSE Consultants, Inc. wrote to the SJRWMD on behalf of the Duda Ranch. BSE stated that they had been working with Duda Ranch since 1988. *"...We were asked to review the adequacy of their existing drainage facilities as well as the effectiveness of their ongoing canal maintenance activities relative to current agricultural land uses...We were also asked to evaluate what impacts their current agricultural operation may have upon future development of lands east of I-95 as well as any other lands contributing runoff through the Duda canal system....*

Our calculations indicated that the culverts were not adequately sized to convey the existing runoff generated from the 25 year rainfall event and we provided recommendations for improvements to those struc-

tures...velocities in the canals often reached erosive levels, resulting in the possibility of the transport of silt and sediments to downstream receiving waters (the St. Johns River)".

BSE provided ranch officials with a listing of sizes and numbers of culverts required in the Two Mile, Four Mile, Six Mile, and Seven Mile canals. While some smaller culverts were recommended for removal, much larger culverts were recommended to replace others throughout the canal system. With wider canals and larger culverts, it was my opinion, which would be confirmed by the subsequent findings of the SJRWMD that velocity would increase throughout the canal system.

The writer with a 5 pound bass caught in 4-mile Duda canal

Specifically, BSE recommended removing a 72-inch culvert, replacing same with two 96-inch culverts; removing another 72-inch culvert, replacing with two 84-inch culverts; removing dual 60-inch culverts, replacing same with three 96-inch culverts, etc. Five other changes, similar in configuration were recommended. I find no fault in any of the BSE recommendations; that was exactly what they were—recommendations. The Duda Ranch considered this massive undertaking as "performing maintenance of their canal systems".

On 22 December 1992, the SJRWMD, by letter to the Duda Ranch addressed issues regarding the 452-acre retention reservoir which was now past due for construction. In this same letter, the SJRWMD pointed out: *"it is apparent the Cocoa Ranch did not receive a construction permit or a "no permit required" letter from the District for the drainage improvements made since 1989....Please amend this application to include construction of any improvements and provide the...detailed description of each improvement".*

During 1993, several letters and/or meetings occurred between the SJRWMD staff, BSE, and Duda Ranch regarding the illegal modifications to the Duda Ranch drainage system, the continuing changes in the design, and the past due construction of the 452-acre retention reservoir. BSE stated that a 14 percent reduction in the peak discharge rate would be realized with the new culvert system and some other changes made in the flow pattern of the peak discharge rate. The SJRWMD disagreed with BSEs findings, indicating their data revealed a 22 percent increase in the peak discharge rate.

In a letter to this writer dated 29 April 1994, from the Permitting office of the SJRWMD, notice was given that the 452-acre retention reservoir and the "after the fact" permit for the drainage of Duda Ranch would be either approved or denied at the SJRWMD Governing Board meeting on 10 May 1994.

I responded by letter dated 5 May 1994 to the Chair of the SJRWMD Governing Board stating I would not be available to attend the 10 May meeting. I requested my letter be read into the record of the meeting; however, my letter was not read into the minutes. I was informed the SJRWMD Governing Board granted permits for the 452-acre reservoir and the "after the fact" permit. I called the SJRWMD and spoke to legal council, Clare Gray. I inquired about the time remaining for SAVE to file for an Administrative Hearing. I was informed the time period elapsed the day of the meeting.

Ms. Gray stated the 14-day calendar began at the time notice was printed in the newspaper of intended action by the governing board, or from the written notice to interested parties, whichever occurs first. One would think a more reasonable rule would allow 14 days following any decision by the Governing Board. This would allow a reasonable time period for any person or group to appeal the board's decision. Had the board denied the permit application, SAVE would not have reason to file an appeal.

In this writer's opinion, the rule is not fairly stated for either party. Any filing would have had to be "ready" and presented almost immediately following a board decision. In fairness to all parties the SJRWMD should change the rule to provide a 14-day "Notice of Action Taken" rather than a 14-day "Notice of Intended Action".

As stated earlier, the SJRWMD had determined the non-permitted actions of the Duda Ranch failed to meet state water quality standards. Unknown to this writer, in May 1993, Duda Ranch entered into an amended consent agreement with the SJRWMD which required several improvements be implemented in the "pumped areas". Also, ranch officials were required to demonstrate that the drainage improvements were consistent with the SJRWMD rule criteria and reimburse the SJRWMD for investigative actions and attorney fees.

The amended consent order did not remove fines of up to $10,000 per offense (or violation).

One is left to wonder, how much investigating was performed and how much legal work was conducted to arrive at only a $1,093 fine. Indeed, this was a most favorable settlement for Duda Ranch for over three years of unauthorized modifications to their massive drainage system and subsequent environmental damage to the water quality of Lake Winder and the St. Johns River.

SAVEs Request for Public Records; Seeks Federal Investigation

As a result of the "unknown to SAVE" amended consent agreement in May 1993 between the SJRWMD and Duda Ranch, and concerns with the location of the 452-acre reservoir in existing wetlands, this writer dispatched a letter to the SJRWMD dated 15 July 1994. Under the Freedom of Information Act, I requested a copy of the complete file regarding all communications with the Duda Ranch, including the amended consent agreement with Duda Ranch. I received a response letter dated July 29, 1994. Patricia Schultz, SJRWMD Central Files, informed this writer that cost for the research and copying would be $300.00.

I made arrangements to review the complete file at the SJRWMD office in Palatka. Accompanied by Mr. Larry Gleason, Vice-President of SAVE, we spent an entire day reviewing the file. We identified the communications we desired, and was supplied with the same. In subsequent review of the records we obtained, we continued to pursue issues relating to the drainage, wetlands, and location of the retention reservoir.

This writer wrote a letter, dated September 15, 1994 to Mr. John Hankinson, Regional Administrator, U.S. Environmental Protection Agency (EPA), in Atlanta, Georgia. In my letter, I requested a federal investigation of the unauthorized drainage of 25,000 acres of gravity flow lands and the impact to significant wetlands on the Duda Ranch.

Mr. Hankinson responded by letter dated October 12, 1994. Mr. Hankinson recused himself since he was a former employee of the SJRWMD.

He appointed Mr. Pat Tobin, Deputy Regional Administrator, EPA, to review my "comments" and involve other appropriate EPA staff to respond to my concerns. One week later, on October 19, Mr. Tobin contacted this writer, stating he was pursuing an investigation of the issues I had raised.

While awaiting further action by the EPA, this writer received a letter from the state Department of Environmental Protection (DEP), formerly the DER, dated November 4, 1994. In the letter, I was informed: *"It may be that the SJRWMD was not as aggressive as they should have been in detecting the original violations and requiring corrective measures. However, my review of the matter leads me to conclude that the Department has no legal basis to object to SJRWMDs resolution of the violations...the review time is limited by law, and once the time has expired there is generally no practical way for the Department to attempt to overturn a water management district action".*

SAVE was hopeful for cancellation of the "after the fact permit" approved by the SJRWMD for the three-year violations of the 25,000 acre drainage basin, in addition to a permit for the 452-acre retention reservoir on lands SAVE considered to be sovereign (public) lands.

Mr. Tobin, EPA, Atlanta office again responded by letter to this writer, dated 23 December 1994. The findings of EPAs investigation were indeed, disappointing. Mr. Tobin's staff had discussed the 452-acre wet detention reservoir and the "after the fact" permit with the Ms. Irene Sadowsky, Corps of Engineers *(COE)* office in Merritt Island, and Mr. Dave Mallard, of the Natural Resource Conservation Service (NRCS) in Brevard County.

Mr. Tobin's letter read, in part: *"EPA and the COE initially believed that a Section 404 permit was necessary to construct the wet-detention system in addition to the permit issued by the SJRWMD. However, the NRCS in November1992 declared the site to be prior-converted wetlands....Prior-converted wetlands are cropped wetlands that, prior to December 23, 1985 were drained or otherwise manipulated for the purpose of, or having the effect of, making a commodity crop possible....Prior-converted wetlands are not regulated under Section 404 of the Clean Water Act and thus are not subject to Section 404 permitting by the COE".*

On the prior-converted wetlands matter, the COE informed this writer that Brevard County Government had approved the conversion of these wetlands to "cropped wetlands" prior to 23 December 1985. Consequently, as stated by the EPA, no authority existed to consider an enforcement action under the Clean Water Act. Therefore, the permits were considered to be valid. Having exhausted my efforts to involve state and federal agencies to remedy the illegal drainage and loss of wetlands inside the dikes of the Duda Ranch, I would now turn my attention to the sovereign lands issue on the expansive Duda Ranch.

Viera—a Development of Regional Impact (DRI)

The East Central Florida Regional Planning Council (ECFRPC) approved SAVE St. Johns River, Inc. to review and comment on the Viera DRI. This writer's initial comments were documented in a letter to ECFRPC dated 27 December 1993. In my letter, I requested a reply on eight concerns. The following concerns are extracted from my letter:

"1) Density levels for residential sites are too high (1/8 to 1/3 acre sites)....

2) Floodplains...will be impacted extensively from its pre-development configuration. Adverse effect on surface waters of the St. Johns River will result. Also, a statewide floodplain protection policy is due release in October 1994 which may impact the floodplain and marsh with more strict controls for development. While the peak discharge rates may not be permitted to exceed present discharge rates, the discharge will occur for longer periods of time and will contain pollutants damaging to the St. Johns River.

3) Proposed built-out is not consistent with SJRWMD policy of restoration of the river system. At a recent public meeting at Viera, provided by SJRWMD staff, the district indicated a desire to acquire some portions of the Duda property near the Ordinary High Water Line (OHWL) for preservation and restoration. Also, action is underway by SAVE...to obtain an OHWL survey by the DEP to determine proper ownership of a significant number of acres of potential sovereign lands. While this item is not a part of the present "plans", it is mentioned since the present "plans" referred to total build-out plans.

4) Total drainage from the proposed project is considered to be excessive (cubic feet per second discharge rates—total volume peak discharge 25 year storm). Our consultant will address this item in more detail—and may not be limited to the 25 year storm.

5) Stormwater retention system will not function properly due to the significant amount of water that must be treated. This same approach was at issue with the proposed Sabal Hammocks project. Game and Freshwater Fish Commission's position (Sabal Hammocks) was that the system would fail within several years.

6) *<u>The combined residential/commercial/golf course, etc</u> planned for this project is simply too excessive for proper protection of the natural resources of the St. Johns River.*
7) *Under no circumstances should any land below the 16-foot contour be considered for mitigation in order to destroy other wetlands on the property. As stated before, these lands must be surveyed to determine rightful ownership...lands that are at 16-foot or below should be removed from any consideration for development or mitigation until such time as proper surveys are conducted by the DEP.*
8) *Statement was made that the St. Johns River was not an area of "Special Concern" or an "Outstanding Florida Waterbody". For over a year, SAVE has discussed this matter with local and state officials. A joint effort will be undertaken...to get such a classification on one of Florida's true treasures, the St. Johns River".*

I concluded my letter with the following statement: *"SAVE would like to assure the developer that we are not anti-development. We did not challenge Viera East, the Government Center, or the Marlins baseball stadium. Ultimately, we may not challenge this proposed project if certain conditions can be agreed upon".*

This writer contacted ECFRPC and stated that SAVE had not received any written response regarding our comments. Viera officials were contacted by the ECFRPC and requested to respond to SAVEs comments. Subsequently, I received a letter from Viera attorney Mason Blake, agreeing to meet with SAVE to discuss our issues. SAVE did not desire such a meeting; we held out for a written response to our written comments. Viera officials never responded in writing to SAVEs comments. Little did the Viera officials realize, they had further energized this writer's determination to ensure no development would occur along the shoreline (my back yard) of the St. Johns River.

<u>Brevard County Commission Supports State Survey of Duda Ranch</u>

Viera officials were unaware of this writer's passion for protecting the water quality of the St. Johns River. On 1 November 1994, I addressed the Brevard County Commission. I reiterated the contents of my letter to Chairman Truman Scarborough, dated 31 October 1994. I quote, in part, from my letter to Commissioner Scarborough: *"the Department of Environmental Protection (DEP)...to survey the Duda Ranch...for the location of the Ordinary High Water Line....I am specifically requesting...the survey include the inside area of the 8-mile canal (east to west) including Units 3 and 4 (location of the wet detention reservoir now under construction). Most of this area is below the 15-foot contour line, with the area around Units 3 and 4 below the 14-foot contour line".*

This writer reminded the commission that in March 1993, at my request, the board notified the DEP that Brevard County was opposed to private development on sovereign lands in Brevard County. Following a discussion by the commission, the board voted unanimously (5-0) to support this writer's request for an Ordinary High Water Line survey of the Duda Ranch. The board directed the County Administrator to initiate the necessary action.

On November 4, 1994, in a letter from County Administration Tom Jenkins, to the Department of Environmental Protection, Mr. Jenkins stated: *"On November 1, 1994, the Brevard County Board of County Commissioners voted unanimously to request the Department...delineate the boundary of State sovereign lands in an area adjacent to the St. Johns River in Brevard County. More specifically, the Board is requesting a survey of...a portion of the Duda Ranch west of I-95,...the inside areas of the Eight Mile Canal east to west, including Units 3 and 4, which is the location of a wet detention reservoir now under construction...(from a letter to Commission Chairman Truman Scarborough from Leroy Wright, dated October 31, 1994, copy attached)....*

The Board of County Commissioners has stated its opposition to private development of sovereign lands. The planned use of the property referenced herein has generated considerable interest among the citizens of Brevard County. Your prompt assistance in providing a sovereign land determination in this area will be greatly appreciated. This writer was copied on this letter as was Mason Blake, Attorney for the Viera Company. This action by the Brevard County Commission would ensure no sovereign (public) lands would be developed on the Duda Ranch.

The Florida Today published an article in the Guest Column on December 29, 1994. My article was in response to a recent Guest Column article by Mr. Tom McCarthy of the Viera Company. First, I addressed Mr. McCarthy's statement that Viera lands did not flood during the recent Tropical Storm Gordon. I quote from my article: *"Very simply, the land has been altered with expanded canals, cross ditches and five large-capacity pumps mounted on miles of huge dikes to offload stormwater into the St. Johns River....Fill dirt, several feet in height, was added before the existing structures were built in order to raise elevations above flood levels.*

The most recent alterations, from 1989 through 1991, were performed without a permit from the SJRWMD. I have reviewed historical documents and maps dating back before the canals, ditches, dikes and pumps were put in place.....The fenced area along the shore of Lake Winder is under water at the present time; so, obviously, the fence is located on public land....The OHWL survey will depict the original configuration of the land, prior to any alterations (reference Florida Supreme Court decision of May 1986—American Cyanamid vs. Coastal Petroleum).

Our county government supports the people of Brevard who placed them into office....The public has not been deceived. We are well-informed and will not be misled by development interest groups, particularly McCarthy, who, as he stated, is director of land development for Viera."

In a letter from Mr. Rod Maddox, Bureau of Survey and Mapping, Department of Environmental Protection, dated February 1, 1995, Mr. Maddox responded to Brevard County Administrator Tom Jenkins letter to the department dated November 4, 1994. Mr. Maddox stated that a preliminary review had been conducted regarding sovereignty interest in the proposed reservoir site, shown as Units 3 and 4 at the Duda Ranch. He also stated that the SJRWMD had provided stage duration elevations along the river profile.

Mr. Maddox also stated that: *"Based upon this analysis, it appears that the following elevations can be used to approximate a "safe line" at selected sites along the river: Downstream of Lake Winder...14.2 feet NGVD; Lake Winder, 15.0 feet NGVD; Upstream from Lake Winder, 16.0 NGVD. I am currently acquiring additional existing mapping, and other technical data, for the purpose of assessing the preliminary location of the ordinary high water line as it relates to the retention area. This will also aid in the design of the survey at this site and others along the remaining areas upstream to Lake Washington...we will be submitting a cost estimate to perform the survey. I will keep you apprised of our progress".*

Also, on 1 February 1995, this writer addressed the Brevard County Board of County Commissioners regarding the Viera Project. This writer used the opportunity to comment on the sovereign lands issue. I informed the commission that I was speaking outside the current DRI boundary (initial 5,800-acre development). I wanted to encourage the commission to "look beyond" the present time-frame. I stated:

"if this project is approved, the County is going to have to make sure that it is on the record that there will be absolutely no conditions allowed whereby the original DRI would in any way permit further construction westward of the existing boundary without the Board coming back with results from a state ordinary high water line survey". I reminded the board that DEP had notified the county in two separate letters that the Bureau has

not determined the ordinary high water line along the river in the area adjacent to the Viera development.

Viera Development Order

During March and April 1995, my message of an Ordinary High Water Line survey of the Duda Ranch became very clear to the Brevard County Commission, who supported this writer's view of action required in the Development Order. Brevard County staff recommended to the Board of County commissioners "protective language" in the Viera Development Order. The following is extracted from page 20 of the Development Order:

"29c. Until a comprehensive management plan is established for the St. Johns riverine system or after five years (whichever is less), there shall be a 322 foot building setback line established landward from the western boundary of the Cocoa Ranch or ***should the Ordinary High Water Line (OHWL) be established*** *by the Florida Department of Environmental Protection, subject to the applicant's adjudication rights, including any judicial appeals, then* ***the setback shall be from the OHWL".***

The official record via the Development Order was now established. The ball was now in Viera's court. In this writer's opinion, Viera officials did not desire a state survey of their ranch. Viera's decision regarding the sell of ranch lands westward of the original Development Order would not be rendered for several years. Viera officials were also keenly aware that development of lands lying west of the initial 5,800-acre site would undergo a state OHWL survey before any plans for development could occur.

This writer was provided a "courtesy copy" of the Development Order for future reference. I converted paragraph 29c to memory. SAVE celebrated this enormous victory, looking forward to one of two options available to Viera, either sell the lands in question to the state, or prepare for years of court proceedings to resolve the rightful ownership of the suspect lands.

On 19 September 1995, in a letter from the Division of State Lands, Department of Environmental Protection, to Brevard County Manager Tom Jenkins, a cost-estimate was provided for conducting the OHWL survey of the Duda Ranch. The total for Bureau expenses and contracting expenses was $165,000. The Division of State Lands expects Brevard County to share in the expenses of the survey. This writer suggested the state pay one-half of the cost, with the balance split between the Viera Company and Brevard County. The Viera Company received a copy of the letter. It was now very clear that there was never an OHWL survey performed on the Duda

8-mile Duda canal from Rockledge to Middle River between Lakes Poinsett and Winder

Ranch as had been claimed by the Viera attorney in an earlier county commission meeting.

State Purchases 14,137 Acres of Duda Ranch

On 14 October 1998, the SJRWMD Governing Board entered into an agreement with A. Duda & Sons, Inc. to purchase 14 miles of Duda Ranch waterfront property which extended landward to the 100-year floodplain. The cost of the property (14,137 acres) was $24.8 million ($1,754 per acre). In 1999, the parties (SJRWMD and the Viera Company) closed on the purchase.

The impending OHWL survey was no longer needed. The public will now enjoy the right to use these "now public lands" as long as the river continues to flow in a northerly direction. All citizens, especially anglers, hunters, bird viewing, hikers, nature lovers, etc. can look forward to exploring the wonders of passive recreational opportunities along the majestic St. Johns River on its "lazy" journey past the former Duda Ranch property.

There are no words to express the overwhelming joy this writer experienced when I was notified that the SJRWMD had indeed purchased this environmentally sensitive property. The 14-mile purchase, as viewed from the south to north included: all of South River (between Lake Washington and Lake Winder) truly an unspoiled wilderness; Lake Winder (located between South River and Middle River and accessible only by boat); and a large portion of Middle River—from the north end of Lake winder to the Eight-Mile canal, located at the north entrance off I-95 into Viera (accessible from boat ramps on Lake Poinsett—several miles to the north).

In today's atmosphere of more and more development, it is indeed refreshing to know there are still a few places where one can "get away from it all". The St. Johns River is such a place.

On 12 July 2000, the Land Management Plan for the River Lakes Conservation Area was published by the SJRWMD. The plan consists of 37,088 acres located in Brevard and Osceola counties. Included in the plan is the 14,137-acre purchase of a major portion of the Duda Ranch. The River Lakes Conservation Area is located, beginning at U.S. Highway 192 north to State Road 520, and includes Lakes Washington, Winder, and Poinsett. Appendix 4 provides a view of planned recreational opportunities soon to be opened to the public in the River Lakes Conservation Area.

As the city of Viera continues to expand westward, the Moccasin Island area is now in public ownership. Restoration projects are underway to separate the city of Viera from the St. Johns River.

More Fishing Trip Memories—In the spring of 1990, I guided a newly-wed couple on a scheduled half-day fishing trip. The spokesperson for the couple was the man. The two had just got married the day before our trip. The young man made a point that catching bass was would be nice, but not necessary to make them happy. They looked forward to the "privacy in the back-waters of the river". I understood his point. I proceeded to an area where small schooling bass were located. They could stay busy catching these smaller bass.

Each of my guests began fishing (at least I thought they had). I turned to point to a spot where several small bass were feeding on minnows near the surface. What I saw when I turned toward the couple was a deep embrace, their rods and reels lying on the floor of the boat. The young lady and I made eye contact. She pulled away from their embrace and said to me: *"We just wanted to go for the boat ride and have a little privacy—I hope you understand"*. Indeed, I did understand. This young couple could have cared less about fishing. After an hour passed, I asked if they would like to take a boat ride around the lake and view an alligator or other wildlife. After all, they were too busy to be bothered by fishing.

On the trip around the lake, I stopped in one spot to let them observe a large alligator near the shoreline. Each of them took a quick look, saying the big alligator was an awesome sight. They then returned to what they had been doing all morning, just embracing, kissing and laughing. This was a first in my experience as a "fishing guide". This trip was more of a "honeymoon adventure" for these two newly-weds.

CHAPTER IX—

THE AMERICAN HERITAGE RIVER INITIATIVE

In February 1997, during President Clinton's State of the Union address, the President stated, by Executive Order, he would designate 10 rivers throughout the nation as "American Heritage Rivers". At the time, this writer served on the Board of Directors for the Florida Wildlife Federation (FWF). I served as Regional Director for Central Florida, consisting of eight counties, while I continued to serve as President of SAVE St. Johns River, Inc.

After President Clinton's address to the nation, I decided to devote my total being to ensure the nomination of the St. Johns River for designation as an American Heritage River. My four-year tenure with the FWF had been both challenging and rewarding. The FWF is continually involved in many issues around the state of Florida. Conservation of fisheries and wildlife, protection of wetlands, restoration of the Everglades, and litigation of issues—when necessary, are among the more notable activities of the FWF.

This writer contacted Manley Fuller, President of the FWF and informed him I was resigning my position on the Board of Directors. Following my resignation, my family surprised me with a life membership in the FWF. I treasure the framed "lifetime" member certificate. I am fortunate to remain a part of the FWF, especially the dedicated office staff in Tallahassee.

The Beginning—an American Heritage River Initiative

In response to President Clinton's Executive Order, the Council on Environmental Quality would assume responsibility for the initiative. The council would review all applications and forward recommendations to the

President for the top 10 rivers to be designated. Mr. Ray Clark, Associate Director to the council would be the contact person. Mr. Clark provided this writer with a copy of the National Register. Within the pages of the register, the rules and criteria whereby a river could be nominated were clearly stated.

Securing one of the 10 federal designations would prove to be very challenging and competitive. In order to meet the criteria, the nomination of a river must have wide-based community support. In addition, any river nominated must have historic, cultural, economic, ecological, and recreational opportunities. The initiative was established as a federal grant program; no federal budget would be established. Any member of Congress was given authority to withdraw the designation on any portion of a river within that legislator's district.

The first order of business for this writer was to obtain the support of our own nonprofit organization (SAVE). On 19 August 1997, in a special called meeting, representatives from 21member groups of SAVE voted unanimously to support the nomination of the St. Johns River for designation as an American Heritage River. SAVE approved a Resolution at the meeting in support of the nomination. The SJRWMD agreed to process all written responses and file the application by the federal due date of 16 October 1997.

This writer called upon a great friend, and a friend of the St. Johns River. Mrs. Pat Poole was a council member for the city of Melbourne, and past President of the Brevard County League of Cities. Pat knew every Mayor and City Manager in Brevard County. She immediately went to work and secured resolutions from 14 of the 16 municipalities in the county. No municipality objected to the designation. Cocoa and Titusville stated there was insufficient time to get the item on their agenda for a hearing before the deadline for submittal.

In the late summer of 1997, this writer contacted the Brevard County Commission and the Indian River County Commission. Both commissions placed this writer on their next meeting agendas. I addressed their respective boards regarding the American Heritage Rivers Initiative. At the meetings, following my presentation, both commissions supported the nomination with resolutions. In both counties the vote was 3-2 in favor of the nomination.

It is significant to note that Brevard County Commissioner Helen Voltz opposed the nomination as did Commissioner Randy O'Brien. Commissioner Voltz's action subsequent to the vote was inappropriate and will be addressed in this chapter. Voting in favor of the designation were Brevard County Commissioners Nancy Higgs, Truman Scarborough, and Mark Cook.

Four Cabbage Palms, a natural landmark north of SR 520; a large hawk surveys the area

On 14 September 1997, I successfully obtained unanimous support of the FWF Board of Directors with a resolution in support of the nomination. I provided the previously approved resolutions from the cities, counties, SAVE and the FWF to the SJRWMD in late September. In addition, many citizens, businesses, environmental and conservation groups throughout the St Johns River watershed wrote letters of support of designating the St. Johns River as an American Heritage River.

Support was overwhelming, but not everyone agreed with the designation initiative. Despite language in the National Register that guaranteed "no new federal regulations", some private property rights advocates attacked the initiative. I faced some of these people during my appearances before the Indian River and Brevard County commissions. Their statements were really "out of bound". Such words as "the program is part of a United Nations plot to swipe sovereign lands from the United States", or "the federal government will be taking over our rivers" were typical among a minority who wanted nothing to do with the federally sponsored program.

State Representative George Albright, District 24 in the Ocala area, informed the SJRWMD by letter dated 24 June 1997 that he would *"seek immediate temporary and permanent injunctive relief in state and/or federal court"* if the SJRWMD decided to go forward with participation in the designation process. Congressman Cliff Stearns of Putman County opposed the designation. He argued that since the initiative resulted from a presidential executive order and had not been approved by Congress, it "*stands exposed as illegitimate".*

Congressman Stearns removed his district from participation. To avoid any potential conflict, the SJRWMD responded favorably to Mayor John Delaney, Jacksonville, when he offered to submit the application for this "once in a lifetime" opportunity. He was not only an elected official, but a very prominent Republican who cared about the river's future and had been very active in improving the water quality in the Lower Basin of the river.

Mayor Delaney filed the application on behalf of citizens, municipalities, and numerous conservation groups throughout the river basins. In a press release published in Florida Today on 17 October 1997, the mayor stated: *"The designation will help the rich heritage of the St. Johns and hopefully will make it easier for our community to secure federal funding to protect and maintain a strong river".*

In the final "count" the application contained 586 favorable responses in favor of the designation, while only 83 responses were opposed. The Florida Today published several articles written by this writer in support of the designation. Following one article by this writer on 27 September, Florida Today's chief editor, on 29 September, wrote of this writer's experience on the St. Johns River and encouraged county officials to join the 14 municipalities in supporting the nomination.

Nation-wide, a total of 126 rivers were nominated. Each river community was hopeful of securing one of the 10 rivers to receive the special designation. Due to the enormous number of rivers nominated, several river communities were judged to have equal ratings. As a result of the close competition, President Clinton would name 14 rivers as American Heritage Rivers. The volume of applications resulted in rescheduling the time-frame to name the winning designations to the spring of 1998.

Following the submission of the St. Johns River nomination, this writer worked closely with the SJRWMD, Jacksonville Mayor Delaney's staff, and with Mr. Ray Clark, Associate Director, and Council on Environmental Quality, in Washington. I was aware of the number of responses supporting the designation vs. responses in opposition.

This writer assessed the odds of securing the St. Johns River as an "American Heritage River" (1 chance in 126). I was convinced the historic, economic, cultural, ecological, and recreation data imbedded in our nomination, together with the overwhelming grass-roots support, would result in a win for our river and the citizens of Florida. The St. Johns River was selected as "an American Heritage River". The official announcement by the President had not been decided at this time.

Storm Clouds Appear Regarding Congressional District 15

The chronology of events surrounding the selection of the St. Johns River for this special designation would be clouded by an abrupt "intervention". In January 1998, U.S. Rep. Dave Weldon, District 15, which includes Brevard and Indian River counties, by letter to the council, **removed District 15 area of the St. Johns River from the American Heritage River Initiative.** This writer was unaware Rep. Weldon had taken such action.

In Weldon's letter, he stated the federal designation was not budgeted, and also, there was little support among his constituency for the initiative. This writer was convinced that Rep. Weldon would have needed input from Brevard and/or Indian River County before he would have taken such action. As stated earlier, I was significantly involved and knew "the numbers" of the overwhelming support for the initiative in District 15.

SAVE has over 2,000 supporters throughout District 15. Before I could assemble an emergency meeting of our organization to determine our course of action to counter Rep. Weldon's action, I was contacted by a SAVE supporter. I was asked to contact Commissioner Helen Voltz's office and request a copy of a letter she had written to Rep. Weldon on the subject. Within a day I had a copy of the commissioner's letter.

Commissioner Voltz misled Rep. Weldon. She indicated that several cities that had supported the designation had changed their minds. My course of action was set. I arranged to speak to the Brevard County Commission on 9 February 1999. In addition, I called upon my friend Pat Poole, and requested she contact the 14 municipalities who provided resolutions in support of the designation. I requested she explain why it was necessary to reinstate, in writing, their support of the nomination.

At the meeting, I requested the board reaffirm its support of their resolution, due to the conflict that had arisen over Commissioner Voltz's letter to Rep. Weldon. Again, the board voted 3-2, with Commissioners Voltz and O'Brien in opposition. I spoke directly to Commissioner Voltz regarding her misleading letter to Rep. Weldon. I pointed out that she had written her letter, not only misrepresenting the facts, but that she did so after this commission had voted 3-2 on 30 September 1997 to support the designation with a resolution.

Commissioner Voltz stated that she wrote her letter before the board voted to support the resolution. The commissioner's letter to Rep. Weldon, prepared on County letterhead was dated 19 November 1997, six weeks following the 3-2 vote by the board on 30 September 1997. Following the commission meeting, with the letter in hand, I provided Commissioner Voltz the opportunity to review the date and content of the letter. I heard no audible response as she departed for the parking lot.

On 10 February 1999, the Florida Today published an article regarding the Brevard County Commission's vote to reaffirm its resolution. In the article, Rep. Weldon was credited as indicating: he objected to the project because it was an executive order issued by President Clinton, meaning it did not get congressional approval, and therefore no money was appropriated by Congress for the program.

Rep. Weldon also stated he was worried that projects such as beach renourishment will have their budgets slashed. Rep. Weldon should have been aware that no budget would be established for this initiative. The whole concept is based upon "grant applications".

As a result of my appearance before the Indian River County Commission, the commission's vote was also 3-2 in support of the designation. However, on a previous vote, the commission had voted to oppose the designation. I have no record of any written objections to Rep. Weldon from the two dissenting votes; Caroline Ginn and John Tippin. However, Rep.

Weldon stated in an article published by Florida Today on 17 February 1999 that : "*One of my reasons for not including the St. Johns as an American* Heritage *River a year ago was the objections raised by local elected officials. If their objections have changed, I will be happy to listen to them*".

Dissenting votes are a matter of record, but that is the full extent of their vote. Such minority positions should not influence state or federal officials to bow to their personal whims. As far as any potential confusion by the Indian River County commission's more recent 3-2 vote favoring the designation, that action occurred before the deadline for submittal of our package to Washington and was a part of the record available to Rep. Weldon.

As I contemplated contacting Rep. Weldon, I wanted to ensure him that his constituency, in fact, overwhelmingly supported federal recognition of the St. Johns River as an American Heritage River. During the period of time that the municipalities were reinstating their resolutions, this writer contacted Ray Clark in Washington. I requested a copy of the records he had received from District 15. Mr. Clark replied that the "file" was not separated by areas of the state or districts.

The data I was seeking was located in various parts of the "file". I requested he "overnight" whatever package he had and I explained my reason for the request. The next day I received an 8-inch document which included all responses from the entire St. Johns River watershed. Joyce and I spent two days "bean counting" the support vs. opposition letters/resolutions, etc. as they applied to District 15.

Within Rep. Weldon's district, there were 43 organizations, 14 resolutions, 24 small businesses, 11 user groups, and 6 environmental/conservation groups on record as supporting the designation. Several citizens' letters in the file were in opposition. Of the conservation groups, SAVEs support counted as only 1 favorable response (we have over 2,000 supporters). The same was true of the 14 resolutions from District 15 municipalities; they were considered 14 favorable responses, when in fact, elected officials in these municipalities represented many thousands of citizens.

Pat Poole provided this writer with reinstatement positions/resolutions which were immediately faxed to Washington, D.C. On 15 February 1999, I appeared as Guest Columnist in Florida Today. I quote from the article: "*We are determined to demonstrate to U.S. Rep. Dave Weldon that he acted on misinformation, rather than the facts and will of the people of his 15th Congressional District when he removed Brevard and Indian River counties from the initiative...the record shows Brevard County had the highest combined supporting number of cities, groups and businesses of any county in the entire length of the St. Johns River.*

During the U.S. Congress spring break in 1999, this writer set up an appointment with Rep. Weldon at his Viera Government Center office. I was contacted and informed the congressman had a pressing issue at the Space Center and would not be able to meet with me. I scheduled another time and date to meet with him. Again, that meeting was cancelled at the last minute. After being informed the congressman would have to return to Washington soon and would not be available, I assured his aide that I would reach him in Washington.

This writer compiled a 21-page package and faxed the entire package to his Washington office. I stated in a cover letter that Rep. Weldon must reinstate District 15 into the American Heritage River Initiative within 10 days and confirm the same to this writer. I also informed Rep. Weldon if he failed to do so, every word in the package would appear in the Orlando Sentinel and the Florida Today newspapers. I guaranteed the entire package would be printed as SAVE was prepared to pay the cost of publication.

Within three days, I received a letter from Rep. Weldon, whereby he informed the Council on Environmental Quality that he was reinstating District 15 into the initiative. I contacted our membership, including my friend Pat Poole, and gave them the good news.

President Clinton to Officially Dedicate the "American Heritage Rivers"

Joyce and I had been contacted in the spring of 1998 by Ray Clark. We were invited to appear with President Clinton on the New River near Banner Elk, North Carolina for the announcement of the 14 rivers that would be receiving the "American Heritage River" designation. Mr. Clark stated I had earned the right to be there and was looking forward to meeting the two of us. The New River would be recognized as one of 14 rivers named at the dedication ceremony. We were surprised and humbled by the invitation. Mr. Clark obtained some personal information from me and stated he would be back in touch when a date was set for the dedication. Weeks past as we looked forward to the opportunity to be a part of the celebration.

I had about given up on a call from Mr. Clark. Joyce and I had planned to drive our motor home to the event and spend a few extra days viewing the great scenery of the mountains. Mr. Clark finally called and apologized for the short notice. The President's schedule was adjusted to include the long awaited dedication ceremony.

Mr. Clark informed this writer our airline tickets were available at the Orlando International Airport for an early flight the next morning. Before I could respond, he indicated our "special passes" would be provided upon our arrival at our hotel. The dedication ceremony was scheduled for the next day. At this point, I apologized as I interrupted, but felt he needed to understand "there is a problem". He said: "oh, how can I help you?" I stated that we had a serious fear of flying; that in fact, we do not fly. There was silence on the line for a few seconds. Mr. Clark responded: are you serious? We were very serious.

I informed Mr. Clark of a conversation I had with my brother in the summer of 1997. Joyce and I were in Alaska; traveling in our motor home. We were on a leisure vacation trip for two months. My brother and his wife "flew" to Vancouver to join us on a 7-day cruise through the Inside Passage to Glacier Bay. It was a fantastic experience. Upon our return to Vancouver I asked them both to take care of themselves. *"If either of you were to pass away, we will miss your funeral; we could not drive to North Carolina in less than 10 days"*. My brother knew I was very serious. I believe Mr. Clark now understood there was no way we were going to fly (anywhere).

Joyce and I had looked forward to the Presidential dedication ceremony, but time would not allow us to drive 700 miles and be assembled by 10:00 AM the following morning. Mr. Clark understood that we would not be attending this prestigious event. He stated to this writer that he had recommended our appearance with the President as a token of appreciation for the dedicated work I had performed in securing the designation for the St. Johns River. He also stated that he was looking forward to meeting me after the many telephone conversations we had shared over the past several months.

As this lengthy telephone conversation was about to end, I asked if anyone from the federal government would be visiting Florida to dedicate the St. Johns River. Mr. Clark responded that indeed, the Secretary of the Environmental Protection Agency, Carol Browner, would be in Jacksonville on 30 July 1998 to dedicate the St. Johns River. I assured Mr. Clark that Joyce and I would be present.

At 10:00 AM on 30 July 1998, Secretary Browner opened the ceremony with a question. Looking out over the crowd, she inquired: "where is the SAVE St. Johns River group? Would you please stand? We were seated on the front row of a large gathering on the waterfront of the river. We stood up, four of us: Joyce and I along with Pat Poole and her husband Bill. Secretary Browner informed the crowd that our organization worked tirelessly and were determined to secure this special designation. Her comments brought a round of applause.

Pat Poole, Carol Browner and this writer celebrate designation of St. Johns River as an American Heritage River in Jacksonville

Following the event, we met with her, took a few pictures and reminisced about her days as Florida's Secretary of the Department of Natural Resources, before President Clinton called her to serve as Secretary for the U.S. Environmental Protection Agency. Although the issue with Rep. Weldon was unknown to this writer on 30 July 1998, the St. Johns River was going to receive the federal designation as an American Heritage River. Indeed, it was a significant highlight for this volunteer citizen—determined to make a difference. (As stated earlier, under the heading **"Storm Clouds Appear....** Rep. Weldon would reinstate District 15 into the designation.)

Federal Officials visit the St. Johns River

On 23 January 1999, Florida Today published an article on a visit to the Upper Basin of the St. Johns River by several federal officials, including Ray Clark. The purpose of their "airboat tour" visit was to observe "first hand" the massive water control project underway in the Upper Basin. This same group of officials also visited the Middle and Lower basins of the river to familiarize themselves with the unique situations of each basin along the 310-mile length of the river.

The Upper Basin tour was conducted by Maurice Sterling, who at that time was Director, Division of Project Management for the SJRMWD. A large photo appearing alongside the article captured the airboat as it headed toward the Blue Cypress Lake area, a part of the ongoing Upper Basin Restoration Project.

In the article, Mr. Clark stated: *"We want to find out where the barricades exist".* He further stated that the cooperation between federal, state and local agencies was the most remarkable aspect of the project (referring to the Upper Basin Restoration Project). As a result of the tour, federal officials hope for a better chance of getting federal funds for more river restoration projects. Henry Dean, Executive Director of the SJRWMD stated: *"There aren't any guarantees, but it seems the more they know, the better opportunity we'll have to apply for federal funds".*

Elected officials gain a better prospective of the river's problems by way of "personal visits". For many years, I have escorted numerous officials and media representatives to view the effects on the river's water quality caused by poor drainage systems and agricultural pump discharges of polluted runoff from farms along the Upper Basin of the St. Johns River. In recent years, I have witnessed these same officials change the old systems to one of treating polluted runoff in reservoirs on their own property. In turn, the river's health is slowly improving. However, other identified restoration projects are not funded and perhaps years from being completed.

In late July 1999, the St. Johns River was provided a federal "river navigator" to assist the newly appointed Steering Committee. Barbara Elkus, a 22-year employee of the Environmental Protection Agency in Washington addressed approximately 30 representatives from the county, the SJRWMD and environmental groups at a meeting held in the Brevard County Commission chambers. She made the point that to obtain federal funding, specific data must be included in grant requests, such as how, when, where, and why such funding was justified. With the special federal designation, the doors in Washington should be a little easier to open.

St. Johns River Steering Committee is Organized

Mayor John Delaney was established as chair for the newly formed Steering Committee. He would be responsible for setting up a river-wide organization to work with the new River Navigator. Mayor Delaney contacted this writer by letter, requesting I serve on the Steering Committee. I accepted, and in turn, recommended the mayor contact Brevard County Commissioner Sue Carlson and Indian River County Commissioner Ruth Stanbridge and request each of them to serve on the Steering Committee. Mayor Delaney followed up; both commissioners were willing to serve.

The Steering Committee was represented by a county commissioner from each of the 14 counties along the river. In addition, the SJRWMD, Fish & Wildlife, DEP, East Coast Regional Planning Council, mayors, and several citizens from the different basins joined the membership. The Steering Committee consists of 25 members.

The first meeting by the Steering Committee was directed by Jacksonville Mayor John Delaney. He encouraged the committee *to* concentrate on achievable, non-controversial projects. He stated: *"We need to show some victories, and we need to show that we are not regulators"*. About 50 people attended the initial meeting in Jacksonville. Among the attendees at the initial meeting was one Mr. Ray Clark. Mayor Delaney requested each participant introduce themselves.

After the introductions were completed, Mr. Clark was asked to make an opening statement of behalf of the Council on Environmental Quality. His first words were a statement about this writer. He wanted the group to know that "*Mr. Wright worked tirelessly on the nomination of the St. Johns River...and it is a pleasure to finally meet him*". Indeed, it was a compliment; this volunteer citizen sitting among city mayors, county commissioners, and staff from the various state agencies.

This was a "get acquainted meeting". Organization strategy was discussed in general terms. Ideas were exchanged on types of restoration or improvement projects. The Steering Committee agreed to meet on a quarterly basis at locations throughout the three river basins.

Over the next several meetings from 1999 into 2000, the Steering Committee organized Citizen Advisory Committees and Technical Advisory Committees in all three basins. These basin advisory committees were encouraged to submit ideas and recommendations to the Steering Committee for evaluation. This process would result in several significant accomplishments during the year 2002.

Action was underway to establish a "St. Johns River Eco-Heritage Corridor". In addition, a contract was awarded to Eagle Productions, Inc.

of Orlando for a documentary on the historic, cultural, economic and recreation assets of the St. Johns River. The Steering Committee would review film footage, and recommend changes, as necessary to satisfy all aspects of the river's contribution to its residents and visitors.

This writer takes Tom Lowe's Eagle Productions film crew on a visit of St. John River to shoot scenes for PBS special—River into the New World

The film would require quality content to assure it would qualify for airing on PBS affiliate stations around the country, and Education channels throughout Florida. Rights to the documentary would belong to the Steering Committee at a point in time after national exposure. This writer accompanied Mr. Tom Lowe and his camera crew from Eagle Productions, Inc. on one of their many visits on the river.

Each basin created a "Projects List" with cost estimates and the order in which the respective basins desired to proceed with solicitation of federal funding. Brevard County submitted approximately 35 projects. Of these projects, initially 17 were submitted by this writer. Among the projects identified was participation in the Eco-Heritage Corridor, sediment removal from four lakes in the Upper Basin, park improvements on the river sites, trails, stormwater treatment systems in a number of areas connected to the river, and an Environmental Center.

At the 20 May 2002 Steering Committee meeting, Mayor Delaney,

Jacksonville, discussed his desire for a comprehensive restoration plan for the entire St. Johns River. Committee members were in agreement that an overall plan for the river would be a significant step forward. Over the next year, the Steering Committee's primary focus was to develop a river-wide restoration strategy plan.

Creation of the St. Johns River Alliance, Inc.

Three significant events occurred in 2003 that would have long-term positive effects for the river. The first event: the Seventh Annual St. Johns River Summit held on 13 and 14 January 2003, in the city of Jacksonville. Instead of addressing issue relating only to the Lower Basin, as the previous six summits did, the Seventh annual summit included all three; the Upper, Middle, and Lower Basins.

The second event: In May 2003, the St. Johns River Restoration Working Group released the "St. Johns River Restoration Strategy". This document provides a road map for future action by the soon-to-be established St. Johns River Alliance (SJRA). The third event: the SJRA, established as a nonprofit corporation was formed in the summer of 2003. Since 1998, the SJRA functioned as a Steering Committee for the American Heritage River Initiative.

On the first day of the Seventh Annual St. Johns River Summit held in Jacksonville in January 2003, the summit focused on greenways, trails and recreational issues and the American Heritage River Initiative. The second day concentrated on water quality and water supply issues. The overall theme of the summit was to promote "restoration" of the historic St. Johns River. The reader should understand that the word "restoration" refers to a long-term cooperative effort to restore, enhance and protect the entire 310 miles of the St. Johns River. Partnerships between the SJRA and local, state, and federal agencies will be key to achieving the much needed restoration of the river.

The Seventh Annual St. Johns River Summit successfully encouraged legislative and grassroots champions of the river to stay the course, provided a forum for identifying needs and strategies for a river-wide restoration plan, and heightened public awareness and scientific understanding of issues related to the St. Johns River.

The St. Johns River Restoration Strategy document was published in May 2003 and provides a road map for future action by the SJRA. Release of this document was a result of hard work, over a period of several months, by a team of dedicated county commissioners from throughout the river's reach, citizens, the City of Jacksonville, SJRWMD, and the Florida

Department of Environmental Protection (DEP). This writer represented SAVE St. Johns River, Inc. as a member of the Working Group. These volunteers contributed many, many hours to ensure the "strategy" was correctly presented.

The Restoration Goal of the restoration strategy is: "*to restore the environmental health of the St. Johns River such that it meets or exceeds state water quality standards and results in significant ecological, recreational, historical, cultural, and economic improvements*".

As stated in Section IV of the Restoration Strategy Plan, the SJRA was created following the release of the strategy plan. Established as a 501(c) (3) non-profit corporation, the SJRA would operate with state-funded grants, membership dues, citizen contributions, etc. On an annual basis, the SJRA will prepare a Priority List to assist funding efforts, along with a legislative strategy. The Priority List projects will be submitted to local, regional, state, federal, and private funding organizations. The total cost estimate for all river projects is $4.6 billion. While some projects are underway, a minimum of 20 years seems a more realistic goal for completion of all projects identified in the strategy plan.

The creation of the SJRA (public/private partnership) is indeed a major achievement that demonstrates how government agencies and citizen support groups are committed to protecting the St. Johns River.

SJRA Achieves SWIM Designation for Upper Basin

A great example of the SJRAs ability to work with government agencies was the SWIM (Surface Water Improvement/Management) designation sought for the Upper Basin by this writer for several years.

In the summer of 2004, I addressed the SJRA Board of Directors on the necessity of designating the Upper Basin as a SWIM body. The designation would put the Upper Basin on a "level playing field" with the Middle and Lower Basins, both of which are designated as SWIM bodies, as is the Indian River Lagoon. This issue was referred to the SJRWMD for a response at the next board meeting.

On 20 October 2004, the next scheduled meeting of the SJRA, Mr. Casey Fitzgerald, SJRWMD, and Mr. Brad Thoburn, City of Jacksonville, addressed the SWIM designation issue. I quote from the minutes of the meeting: "*Mr. Fitzgerald noted that there are important reasons why the designation has not been sought and no formal motion by the Board is needed. The Water Management District will be prepared at the next meeting to present the issues surrounding this designation*".

At the SJRA board meeting on 5 January 2005, Mr. Maurice Sterling,

SJRWMD, provided a "power point" presentation regarding "Upper Basin SWIM Designation Analysis". This writer quotes from the minutes of that meeting: *"Mr. Sterling stated that the Upper Basin Restoration Project has received over $100 million dollars in funding for land acquisition and $100 million dollars in funding for structural improvements.... The District feels the lack of designation has not impeded the upper basin in receiving funding and stated that the Indian River Lagoon is a SWIM body and many projects in the Upper Basin benefit the lagoon and vise versa.*

Member Leroy Wright voiced his concerns over the lack of funding, and projects north of U.S. 192...Chair Sue Carlson echoed his concerns. Mr. Sterling voiced several concerns in designating the Upper Basin as a SWIM body; in particular, jeopardizing existing funding sources...Mr. Sterling and Chair Sue Carlson agreed to meet to discuss the issue".

When the SJRA convened on 15 July 2005, I had been placed on the Agenda to speak to the SWIM designation for the Upper Basin. With the knowledge that I had the support of the alliance Chair, Commissioner Sue Carlson, I presented a clearly defined approach to securing the SWIM designation. Regarding Mr. Sterling's statement relative to the millions of dollars being spent for the Upper Basin Restoration Project, which SAVE fully supported, I was quite familiar with the restoration boundaries. I informed the SJRA board of numerous canals that offloaded into the St. Johns River, located north of U.S. 192, the northern boundary of the ongoing Upper Basin Restoration Project.

Stormwater systems were needed to treat runoff north of U.S. 192; from the new city of Viera, Rockledge, Cocoa, and numerous other canals in the Upper Basin located miles north of Melbourne along the river. As a SWIM body, the Upper Basin would receive an appropriate share of the millions of dollars available for stormwater and restoration projects, north of U.S. 192. At this same board meeting, the SJRWMD distributed copies of funding allocations for the Lower and Middle Basins, plus the Indian River Lagoon.

The Lower Basin received approximately $3 million; the Middle Basin received approximately $2 million, while the Upper Basin received no funding. Lower and Middle basin members knew if the Upper Basin was added, the "pie" would be sliced 4 ways, rather than 3 as it has been since inception of the SWIM program. I called for a vote on the matter; the vote was unanimous to include the Upper Basin as a SWIM body.

Thanks to the support of the SJRA, this writer's years of effort was finally approaching a successful conclusion. Within a few days following the unanimous vote, SJRA Executive Director Mindy Matthews notified Chair Sue Carlson and this writer, to be available to present the SJRA position to the SJRWMD Governing Board on 13 September 2005.

This writer fully expected the Governing Board to approve the SWIM designation for the Upper Basin. Joyce and I planned to meet Mindy and Sue in Palatka prior to our scheduled appearance to ensure we had our "ducks in a row".

At midnight before our scheduled departure at 8:00 AM the next morning, I rushed Joyce to the Emergency Room at the local hospital. She was quickly diagnosed with an appendix that was near bursting. Surgery was performed about 2:00 AM. She remained in the hospital for two days. I would not be attending the Governing Board meeting. After notifying Chair Carlson of my situation, I was assured that she and Mindy could handle the issue. When the two of them arrived, they discovered that the SWIM designation had been placed on the Consent Agenda and approved—no discussion was necessary. And Joyce recovered very well.

The SJRWMD is preparing a SWIM plan for the Upper Basin. This writer will be following up on the publication of the plan. The SJRWMD staff has assured this writer that the Upper Basin will receive its share of future funding along with the other SWIM bodies.

More Fishing Trip Memories—The year was 1994; this fishing trip occurred on Lake Okeechobee in south Florida. In January, Joyce and I visited the lake to practice for an upcoming "Guys & Dolls" bass tournament the following week. We located fairly clear water near Horse Island, a 20-minute run from the tournament site. During our practice, each of us caught several "tournament keeper" bass. I was especially pleased using my favorite 7-inch black & blue worm. We were not counting the number of times we "took the worm away" from the bass to avoid any chance of running the fish away from the area.

The following Saturday, along with approximately 110 boats competing in the tournament, we were off to our chosen site, hopeful no other boats would be "sitting" on our spot. We were in the second out of four flights of boats to depart the tournament site. As we ran the 12-mile route to Horse Island at a speed of 55 miles per hour, we were passed by other boats running 65 to 70 miles per hour. Upon arriving, no boats had entered the area of Horse Island.

Joyce was fishing with a red shad colored worm; I chose to fish my favorite black & blue colored worm. Very soon, Joyce had landed 3 nice keeper bass. During that time period, I had only one strike and did not catch the bass. As she placed her third bass into the live-well, she was all smiles and encouraged me to switch to a red shad colored worm. I stated: *"I caught plenty of bass here last week with my black & blue worm—don't worry, I'll catch up with you"*. Soon, she landed her fourth bass. I decided

I would change to a red shad color, not believing it would really make any difference.

On my first cast, I had a tremendous strike. I promptly set the hook and had what I believed was a giant bass on the line. Quickly, the line went limp. The fish had broken free. It broke free alright; when I looked at my hook, it had broken at the bend of the hook. I tied on another red shad worm and caught a nice bass which I promptly put in a separate live well. Out of 110 boats, our catch put us in Ninth place. Oh, if I had only listened to Joyce, we would have done so much better. She is willing to experiment with different lures much more than I do. I should have realized she was on what bass anglers call—"a winning pattern".

In 2004, after my older son Steve relocated to Florida, I took him fishing. Steve had not fished very much over the previous 20 years; I felt the need provide a little advice on bass fishing. Before we started fishing, I pointed out to him the need for patience in fishing plastic worms. I reminded him that if he felt a light "touch" while slowly dragging the worm along the bottom, he should reel in any slack in the line and wait for a second "touch". If he felt a tug or pull on the line, I advised him to promptly set the hook. Again, I reminded Steve that strikes could be few and far between when bass fishing; he would need to learn the art of patience if he was to experience any success in bass fishing.

After the 15-minute discussion period was over, I looked at him and said: *"Now, let's go fishing"*. Steve cast into the edge of the river's current; no more than 5 seconds elasped before he hollered: *"I've got one on"*. He landed a "keeper" before I ever made the first cast. So much for "patience". However, I again reminded him of the other situation he may have to endure, the time that would pass between catches. I suppose one could say the teacher was right on this one; Steve never caught another bass on that trip. I managed to catch several keepers that morning. The whole point in a family relationship is not how many fish you catch, but the quality of family time spent together.

CHAPTER X—

FRUITS OF MY LABOR

This chapter summarizes this writer's "volunteer" work for the protection of the historic St. Johns River. This magnificent river was in "critical" condition. My effort was born out of a special love I have for God's gift to all of us, a river that was being neglected. I was determined to "stop the bleeding". I turned a deaf ear to those who would say: "you cannot stop the polluters, and you will not stop developers".

You, the reader will judge my work and decide if "fruits of my labor" proved my claim that "one person made a difference". All of us have expertise or interest in something we value. I encourage you to pursue your interest, whether in religion, social issues, environmental, etc. If you choose not to become involved, then look at yourself in the mirror—you will see the person who decided not to rock the boat.

I appreciate the recognition and honors awarded me over the years. Indeed I was surprised and humbled by each experience. Being a volunteer is the giving of one's self for a purpose. As the title of my book states, my purpose was: Saving the St. Johns—an American Heritage River. The highlights of my "volunteer" journey are briefly summarized in the following paragraphs.

1985—Founder and President of SAVE St. Johns River, Inc.

I organized a public meeting to seek support in forming a coalition to challenge existing policy and regulations regarding the restoration and protection of the St. Johns River. The coalition was formed at this initial meeting. The coalition grew to 30 groups, consisting of over 3,000 supporters. Our coalition membership evolved from businesses, homeowner associations, fishing clubs, hunting clubs, conservation groups and citizen support members.

The coalition was successful in fundraising events which provided the financial means to pursue our mission. The coalition established a positive reputation among conservation groups and government officials throughout the St. Johns River communities. We celebrated our 21-year anniversary in October 2006.

1986—Established Aquatic Weeds Control Maintenance Schedule

I contacted Mr. Jim McGee, U.S. Army Corps of Engineers Office, Jacksonville District regarding excessive invasion of exotic vegetation in the Upper Basin of the St. Johns River. A short time later, Mr. McGee, accompanied by an assistant met this writer at Lake Poinsett Lodge. By boat, we visited Lakes Poinsett and Winder. Unable to proceed into Lake Winder because of the thick blanket of hydrilla, Mr. McGee promised action by the Corps of Engineers. Within the next two weeks, the lakes were treated to kill the hydrilla.

In a letter to this writer dated 17 March 1986, Gail G. Gren, Chief, Construction-Operations Division of the COE extending her appreciation in ""*SAVE St. Johns River designing the hydrilla control operation in Lakes Winder and Poinsett for the year"*. A maintenance schedule to treat the exotic vegetation in the Upper Basin was implemented.

Navigation problems no longer exist; however, some anglers, including this writer are now frustrated. Excessive treatment has included every "green" plant visible to those contractors spraying to remove hydrilla. Moreover, when surface winds are blowing at 15 to 25 miles-per-hour, spraying operations should not be performed. The Upper basin fishery has been scattered as only minimal aquatic vegetation exists to provide cover and habitat for bait-fish and largemouth bass.

1987—New Boat Ramp at James Bourbeau County Park, SR 520

This writer worked with former Brevard County Commissioner Sue Schmitt to replace the old and damaged single lane boat ramp at James Bourbeau County Park with a new "floating" structure. I provided Commissioner Schmitt a drawing, depicting the slope and configuration for a floating dock system. The width of the proposed boat ramp would provide for launching of two boats at a time, complimented by docks on each side and a third dock in the center of the new boat ramp. Within several months, the new boat ramp was built. Commissioner Schmitt had a large sign erected, dedicating the new facility to SAVE.

1988 to Present—Restoring Lakes Hell N' Blazes and Sawgrass

As discussed in Chapter IV, SAVE has been involved in restoration of these lakes for 15 years. In 2003, funding was approved for restoration of these lakes. Subsequently, all COE Project 206 funding was halted by the Corps of Engineers. To complicate matters, once again the cost of restoration by the Corps of Engineers escalated to $21 million dollars. I will continue to seek other funding sources to secure the restoration of these lakes. The mission is two-fold; protect the Lake Washington potable water supply for 150,000 residents in Brevard County and restore the once great fishery of these two lakes.

1989 to Present—Proposed Sabal Hammocks Project

Chapter V addresses this issue. I claim victory on behalf of thousands of citizens. In my opinion, Judge Bruce Jacobus failed the people in his ruling against the state of Florida. The landowner's own survey conducted in 1977 placed the OHWL at 13.0 ft elevation The state survey defended the OHWL at 13.0 ft elevation. At the trial, the state revealed that the 18,000 ft dike on the property was built at the 12.1 ft. elevation. Judge Jacobus declared 12.1 ft. to be the OHWL.

This writer claims victory for the people based upon the fact the landowner is now willing to sell the 720-acre tract. I am working with local agencies, funded to purchase such environmentally sensitive lands to arrange a visit to the site and work out an arrangement with the owner for the purchase of these endangered lands and remove the illegally constructed dikes, allowing the waters of the St. Johns River to reclaim the historic lake-bed of Lake Poinsett.

1990—Adopt A River Project

After a year-long delay, SAVE was approved by the SJRWMD and Brevard County Government to adopt the Upper Basin. The SJRWMD built large signs which read: ADOPT A RIVER—SAVE St. Johns River, Inc—Do Not Litter. With assistance from the SJRWMD and SAVE members, signs were placed at all county boat ramps, private marinas, and fish camps throughout the Upper Basin of the River. SAVE members, along with many citizens remove trash and debris from the waterway each time we visit the river. An added benefit, of course, SAVEs signage reminds all boaters to keep the waterway clean.

SAVE member Scooter Creel assists this writer in placing an Adopt A River sign at James Bourbeau Memorial Park off SR 520

1991—Duran Water Management Plan for Duda Ranch

I received an alert from one of SAVEs networking groups concerning an upcoming meeting in Titusville (before the Brevard County legislative delegation) and a meeting in Viera (before the Brevard County Commission) on the same subject and at the same time of day. I was informed the meetings involved a private water management plan for the Duda Ranch. I contacted each representative of SAVEs 25 membership organizations; I assigned members to attend each meeting. SAVE was well represented at both locations. SAVE was prepared to request each board table the matter, until a state Ordinary High Water Line Survey could be performed to determine if lands involved were state-owned sovereign lands.

When the teams representing the plan realized both meetings were being attended by a large number of SAVE members, the item was withdrawn within minutes from both locations. As one of our network partners stated: *"We picked up our briefcases, smiled at everyone, and left"*.

1991—Floodplain Protection Policy, all Five Water Management Districts

SAVE appeared before the Governor and Cabinet in late 1991. During arguments about floodplain encroachment by developers of the proposed Sabal Hammocks Project, the state ordered all five water management districts to develop consistent and coordinated plans to address floodplain protection policy throughout the state. The five water management districts complied with the order, developing consistent floodplain protection policies within their respective districts during the following year.

Fall of 1991—Bassmasters Support SAVE Position in Proposed Sabal Hammocks Project

In the late summer of 1991, I obtained the support of Bassmasters, in opposing the proposed project named above. I worked with Mr. Al Mills, Corporate Director of Natural Resources. The corporation and its national magazine specializes in promoting bass fishing as well as environmental protection and conservation all across the United States. Mr. Mills wrote a two-page letter to Governor Lawton Chiles, wherein he expressed Bassmasters opposition to the proposed Sabal Hammocks Project. As a national publication, I believe Bassmasters played a significant role in the ultimate vote by the Governor and Cabinet (7 to 0) which propelled the case into court.

Fall and Winter of 1991—Tosohatchee State Reserve

In November, concerned citizens in Brevard and Orange counties contacted this writer regarding potential closing of the Tosohatchee State Reserve. The 28,000-acre Reserve borders 19 miles of shoreline of the St. Johns River, north of SR 520 in Orange County. After meeting the citizens at the park, we toured the area. The Reserve is a beautiful and pristine wilderness with a looped drive-thru dirt road and hiking trails. The Reserve is a popular area for riding horses and viewing the river and wildlife from several locations.

On 22 December, I wrote a letter to Fran Mainella, Director for the Division of Recreation and Parks in Tallahassee, stating SAVEs opposition to closing the Reserve. After several discussions with Mrs. Mainella and her staff, the state agreed to leave the Reserve open. I was happy to inform the park advocates they would retain the use of the park. The issue was settled much sooner than I had expected.

1992—Bassmasters Story: Adverse Impacts of Agricultural Pumping Operations

As a member of Bassmasters for many years, I again requested the assistance of Al Mills, Director of Natural Resources for Bassmasters. I requested he dispatch Mr. Robert Montgomery, Senior Writer with Bassmasters to meet with this writer for a significant story of a dying fishery in the Upper Basin of the St. Johns River. In the December 1992 issue of Bassmasters, Mr. Montgomery articulately described his findings, complete with diagrams of how agricultural pumps functioned to benefit the agricultural interests while adversely degrading water quality in the Upper Basin of the river.

1993—Conservationist of the Year, Florida Wildlife Federation

On 15 April 1993, SAVE Vice-President Mr. Larry Gleason, forwarded a three-page letter to the Florida Wildlife Federation nominating this writer to receive the Florida Wildlife Federation's "Conservationist of the Year" award. Statewide, this prestigious award is highly competitive. Federation President Manley Fuller informed this writer that I had been selected to receive this prestigious award at the federation's annual meeting at Greneleaf Resort in Haines City, Florida on 6 November. My entire family and many friends joined Joyce and me for the event. I was grateful, yet humbled, especially for the support of so many friends, associates, and government officials.

1994 to 1998—Regional Director, Florida Wildlife Federation

An honor indeed was the opportunity to serve on the Board of Directors for the Florida Wildlife Federation. All board members view conservation of our woods and waters the highest of our priorities. I resigned my position to focus on securing the designation of the St. Johns River as an

American Heritage River. Anyone seeking to volunteer in support of a great statewide group of 14,000 members should consider joining the Florida Wildlife Federation. I speak from knowledge of their mission; dedicated to protecting and preserving the natural resources of the great state of Florida.

1994—Elevated Spans on New Bridges over St. Johns, SR 520 and US 192

In January, Mr. Bill Daniel of Oviedo, addressed the Board of Directors, Florida Wildlife Federation. Mr. Daniels presented a package of data in support of a raised bridge span over the St. Johns River at SR 520. Bill was a member of SAVEs affiliate club, St. Johns River Valley Airboat Association.

On 26 January, in a letter to Manley Fuller, President, Florida Wildlife Federation, I requested he contact Governor Chiles. I pointed out several advantages for a raised span over the river. The raised span would improve access for larger boats; wildlife habitat would be protected by crossing under the bridge span along the shoreline of the river; and reduction in motor vehicle accidents via a clear view of traffic. On 15 February, the federation dispatched a letter to Governor Chiles supporting SAVEs position.

On 1 June, I wrote a letter to Mr. Frederick Birnie, District Environmental Manager for the Department of Transportation (DOT). I strongly stated the need for the elevated span for the bridge. Mr. Birnie responded to my letter stating a public meeting would be held prior to any final decision. Following the public comment meeting, the project was approved for construction. A win for improved public access, safety, and wildlife protection.

Ironically, twenty miles south of SR 520, the US 192 bridge over the St. Johns River was also scheduled to be replaced. Plans called for an elevated span over this bridge. The reader may recall the public sponsored the DOT plan. Chapter IV describes how Mr. Chase, aide to US Representative Dave Weldon attempted to derail the high span bridge. This writer addressed the issue in support of the high span. The DOT proceeded with construction of the high span bridge. This was a significant win for all boaters who can readily boat from Lake Washington to Lake Hell N' Blazes.

1995—Land Conservationist of the Year (Nomination) Florida Wildlife Federation

This nomination, although I was not selected to receive the award, demonstrates an example of the support I received from Brevard county government officials. On 25 April 1995, Lisa Barr, Director of Natural Resources, wrote a letter to the Florida Wildlife Federation, nominating this writer for this award.

Lisa stated: *"Mr. Wright has emerged as one of the leading advocates of preserving and restoring the St. Johns River. His love and respect for this river were evident as he vigorously worked toward its protection during the Viera DRI".*

In Lisa's closing statement, she quoted Martin Luther King, Jr. who once said: *"the ultimate measure of a man is not where he stands in moments of comfort and convenience, but where he stands at time of challenge and controversy"*. Lisa concluded her letter with these words: *"I believe Mr. Wright's stand during this very controversial time is a tribute to his character and merits consideration for Land Conservationist of the Year"*. Thank you Lisa for your kind words and support.

1995 to 1999—State Purchase of 14,137 acres of Duda Ranch

In 1995, I was in a unique position, having been granted reviewer status of the proposed and planned new city of Viera. Chapter VIII provides the details of how this writer achieved local and state agency support in the purchase of 14,137 acres of the Duda Ranch. The property is now being restored as a passive recreation area as part of the River Lakes Conservation Area. This lengthy four-year process resulted in a positive and significant win for public access to the historic hammocks and floodplain of the St. Johns River, while restoring thousands of acres of marsh.

1995: 10-Year Volunteer Service Award

The SJRWMD presented the award to this writer in a ceremony held at the district's headquarters in Palatka. It was indeed an honor to be recognized by my peers and associates whom I had worked with to improve and restore the St. Johns River since 1985. My volunteer services continue to the present day.

1996 to Present—Annual Cleanup of the St. Johns River

In 1996, the SJRWMD, supported by businesses and citizen volunteers sponsored the first ever river wide cleanup of the entire St. Johns River. The event occurred simultaneously throughout the watershed on 2 March. The cleanup was a tremendous success. I served as Chairman for Upper Basin's Participation in the initial event. I have served as a Site Captain for eight years.

Individually, there are too many sponsors and participants to list; however I must point out that Keep Brevard Beautiful played a significant role in making the effort so successful in the Upper Basin. The annual cleanups continue; however, due to logistics, each basin (Upper, Middle, and Lower) now schedule their own time and dates for the cleanup. The event is a great opportunity to show support for a clean river.

1996—Building Weather Shelters Along the Upper Basin of St. Johns River

In early 1996, I became aware of the SJRWMDs plan to remove all the old camps along the river. Reasons given were the absence of liability insurance on these camps and their potential for causing environmental damage to the river's water quality. I became involved as I had taken shelter in one or two of these camps during storm events when it was not possible to make it back to the boat ramp, miles away.

This writer served on a committeee that recommended location of weather shelters in the Upper Basin of the river. I opposed the removal of any existing camps until the weather shelters were actually constructed. Today, there are numerous shelters, appropriately located, that serve the public's need for protection during these storm events. A great example of the SJRWMDs desire for public safety, while protecting the river._

1996: Proposed Exchange of 10,000 Acres of Bull Creek Wildlife Management Area

Friends of Bull Creek was formed as a result of several conversations I had with members of a hunting group who utilized Bull Creek for its popular hunting opportunities. I encouraged Mr. Don Aycock, with the support of others in the hunting group to form an organization. I also stated SAVE would be willing to accept their group as an alliance of SAVE as long as conservation and public access to state lands were a part of their mission.

Friends of Bull Creek joined SAVE in 1996. SAVE supported the group with numerous communications with the SJRWMD regarding lands being considered as a trade for 8,800 acres of marsh in the St. Johns Upper

Basin. Ultimately, the 10,000 acres would remain a part of the 23,000-plus acre Bull Creek Wildlife Management Area.

1997-1998: St. Johns River Awarded "American Heritage River" Designation

Chapter IX reveals the extent of my personal commitment to secure federal recognition of the St. Johns River as an American Heritage River. Receiving this special designation has to be one of the more significant achievements this writer has been a part of over the past 21 years. I am especially thankful for the opportunity to serve with a great group of individuals on the Board of Directors for the St. Johns River Alliance, Inc. The mission of the alliance can be stated in several words: "to preserve, restore, and protect the St. Johns River".

1998: Conservation Organization of theYear

On 8 April 1998, Mr. David Cox, Marine Biologist with the Florida Game & Freshwater Fish Commission forwarded a letter to the Florida Wildlife Federation, nominating SAVE St. Johns River, Inc. as Conservation Organization of the Year—1998. Dave's letter addressed the many accomplishments of our coalition since the group organized in 1985. The Florida Wildlife Federation, by letter, dated 15 July 1998, confirmed this writer was selected to receive this prestigious award on behalf of SAVE. Thanks to my good friend Dave and my wife Joyce, for their continuing support of this writer's activities related to restoring the St. Johns River.

March 2004: Resolution Presented *by Brevard County Commission*

Brevard County Commissioner Sue Carlson contacted this writer in late February relative to SAVEs project list for the Upper Basin. She requested I appear at the commission meeting on 2 March and report on the issues SAVE was pursuing. She stated that she would allow me to speak under the "Reports" portion of the meeting which was early on the agenda.

Joyce and I were present for the 9AM start of the meeting. I was prepared to provide a status of our river projects. As I began to speak, Commissioner Carlson interrupted me. She stated that if I wanted to address the projects, she would allow me time to do so, but that was not the primary reason I was here.

Commissioner Carlson presented this writer with a framed Resolution. She read the contents of the resolution, and as is customary on such occasions, posed with me for a photo. I was pleasantly surprised by the action of the board. Commissioner Carlson stated to me that space would not permit listing all the good things I had accomplished for the river. She said, we believe the most significant accomplishments are included. The following is an extract from the Resolution:

1) Conservation and restoration of the Upper Basin;
2) Due to his commitment to preservation of the historic, environmental, economic, and recreational qualities, the St. Johns River was designated An American Heritage River;
3) Instrumental in the acquisition of over 14,000 acres of land…for future generation;
4) Founder of SAVE St. Johns River, Inc., served as Director, Florida Wildlife Federation, a former member of the American Heritage River Steering Committee, and currently a board member of the St. Johns River Alliance;
5) Led the "charge" for restoration of Lakes Hell N' Blazes and Sawgrass;
6) Initiated the first annual St. Johns River Clean-Up in Brevard County;
7) Conservationist of the Year—1993, Florida Wildlife Federation;
8) Conservation Organization of the Year—1998, Florida Wildlife Federation.

2004: LEROY WRIGHT Recreation Area

Wednesday, 6 October 2004—a very special day I will always remember. Joyce found a way to get me to the river without my boat. She asked Margarita Engel of the SJRWMD to call me and request I accompany SJRWMD staff to view a river project I had submitted, located north of SR 520. I received Margarita's call on 5 October and agreed to the trip. Joyce and I were to meet SJRWMD staff at 10 AM the next morning at a new county recreation facility off SR 520 at the river. As a light rain was falling on Wednesday morning, I mentioned to Joyce that a little rain would not deter the SJRWMD staff. I retrieved our rain-suits from my boat before departing our home.

Having recently undergone eye surgery, my vision was somewhat blurred. As we approached the park, I was unaware of a tent located to my left. Driving to the dock area, the airboat had not arrived. I said to Joyce, *"I suppose they are running a little late"*. She turned the vehicle into a nearby parking space. Almost immediately, someone was opening the door on the passenger side of our vehicle. It was Commissioner Carlson. Somewhat surprised, I asked if she was going with us on the tour. Her response was: *"No, and you aren't either"*. She asked Joyce and me to come with her. We walked a short distance and stopped; I could hear low sounding voices, but could not identify anyone in the crowd near the tent.

A tall sign had been installed; it was covered with a canvas tarp. Commissioner Carlson requested Joyce to stand on one side of the covered sign and I on the other side. At this point, I knew something special was about to happen—and somehow I was to be a part of it. On a count of three, Joyce and I pulled on cords which dropped the tarp to the ground. I walked to the front of the sign. As I was now facing the sign, I realized why I was here. In very large letters (thank God) the sign read: **LEROY WRIGHT Recreation Area**.

Dedication Ceremony of new park; shown in photo are on left, my daughter Donna, this writer; on right side my wife Joyce, sons Robbie and Steve

Overwhelmed, I could hardly speak for a couple of minutes. At the time, I did not recognize my family was present, including my daughter, Donna, and son-in-law, Mike, from Georgia. As each person approached me, offering congratulations, I could finally determine who they were. In addition to my wife, Joyce, and daughter, my sons Steve and Robbie were present. Representatives from Florida Fish and Wildlife Conservation Commission, County government, state officials, Keep Brevard Beautiful, Florida Today media, SAVE members, and friends were present for this very special event in my life.

It was obvious to everyone; I never had any indication of how this came about. I only assumed it had to be very recent. I was wrong. I now know that Joyce initiated this effort in January 2000, when the park was in the very early stages of development. Numerous letters from government officials and conservation groups had responded very positively to Joyce's initial letter requesting support for naming the park in my honor.

Brevard County Parks and Recreation Assistant Director Jack Masson had to write a "new" policy whereby someone still living could be honored in such a special way. Jack is a good friend I have known for over 30 years. In my conservation with him at the dedication ceremony, he informed me that Joyce began the effort four years earlier. He continued our conversation in his "dry sense of humor" as he stated: *"Leroy, if you had just passed on, this would have happened years ago"*. Now, there's a real friend. Again, what an honor to be recognized in such a special and permanent manner.

After Dedication Ceremony, left to right—Comm. Sue Carlson, Parks & Rec Asst Director Jack Masson, this writer and Joyce

2005—"SWIM" Designation Awarded to Upper Basin of St Johns River

Chapter IX provides the details of this significant accomplishment. This writer continued in pursuit of obtaining the SWIM (Surface Water Improvement and Management) designation for the Upper Basin for a number of years. In earlier meetings of the St. Johns River Alliance, I had obtained a unanimous vote of the board to recommend the SJRWMD include the Upper Basin in the program.

The opportunity to receive state funding for projects in the Upper Basin (north of U.S. Highway 192) near Melbourne, would be greatly enhanced with the SJRWMD Governing Board approval of the designation. St. Johns River Alliance Chair, Sue Carlson, Executive Director Mindy Matthews and this writer were scheduled to appear before the SJRWMD Governing Board at the October 2005 meeting. However, no discussion was necessary. The item had been placed on the Consent Agenda, and was approved by the SJRWMD Governing Board. A very significant accomplishment; my personal thanks to the St. Johns River Alliance board members for their support.

End of Story

It was interesting and enjoyable, revisiting the past 21years. There were many more successes than disappointments. Failure is not a part of my vocabulary. Disappointments are not failures; such projects or issues simply require additional time and effort. Many citizens volunteer a lot of their time for some great causes. Keep up the good work. Stay focused; never accept negative comments or findings from well-intended friends or government officials as the final decision. Determine for yourself when it is time to move on to another venture.

My faith sustains me in all matters of life. I give honor and praise to my creator, the God of the Universe who sacrificed his Son, Jesus, in whom I believe. God gave man dominion over his creation.

We must protect and preserve these gifts from our creator. God bless each of you.

APPENDIX 1 - UPPER BASIN BOUNDARY MAP - ST. JOHNS RIVER

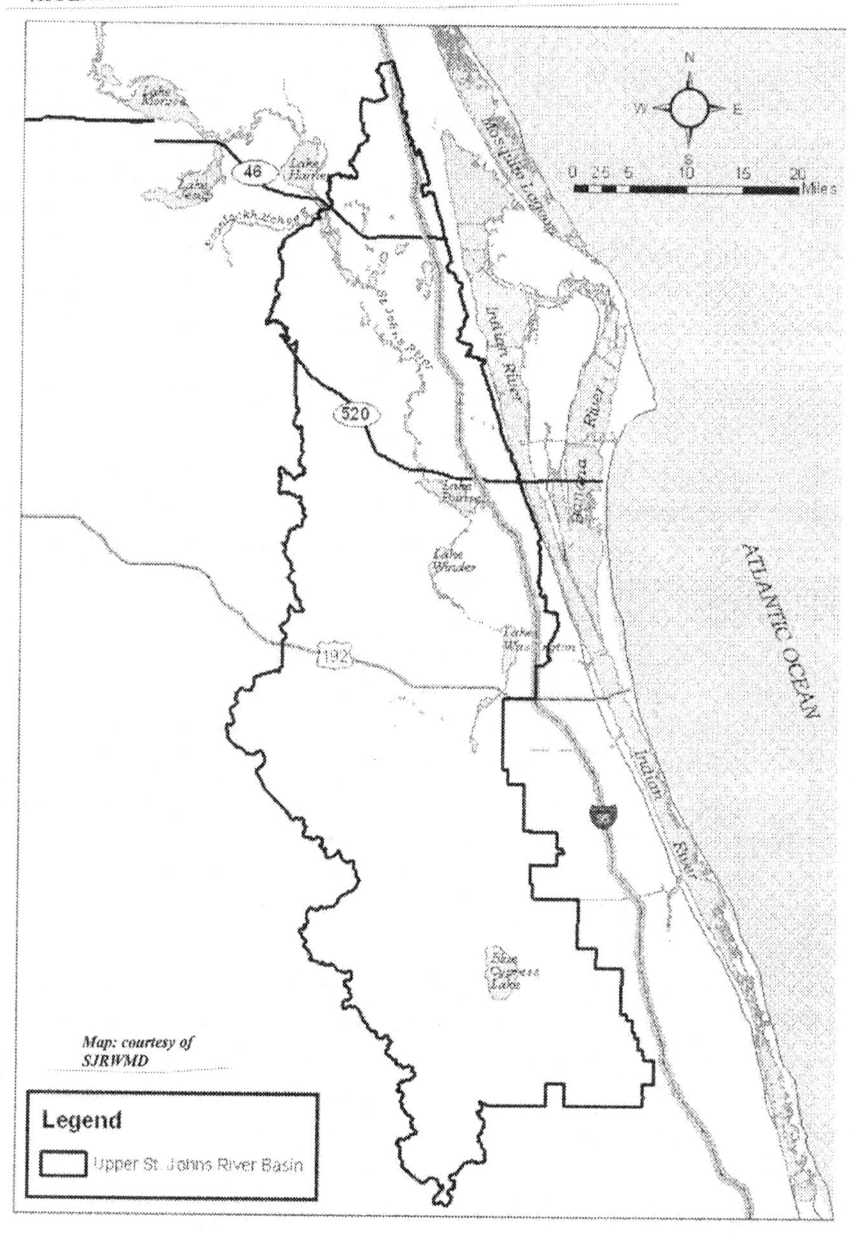

APPENDIX 2 - THREE BASINS' BOUNDARY MAP - ST. JOHNS RIVER

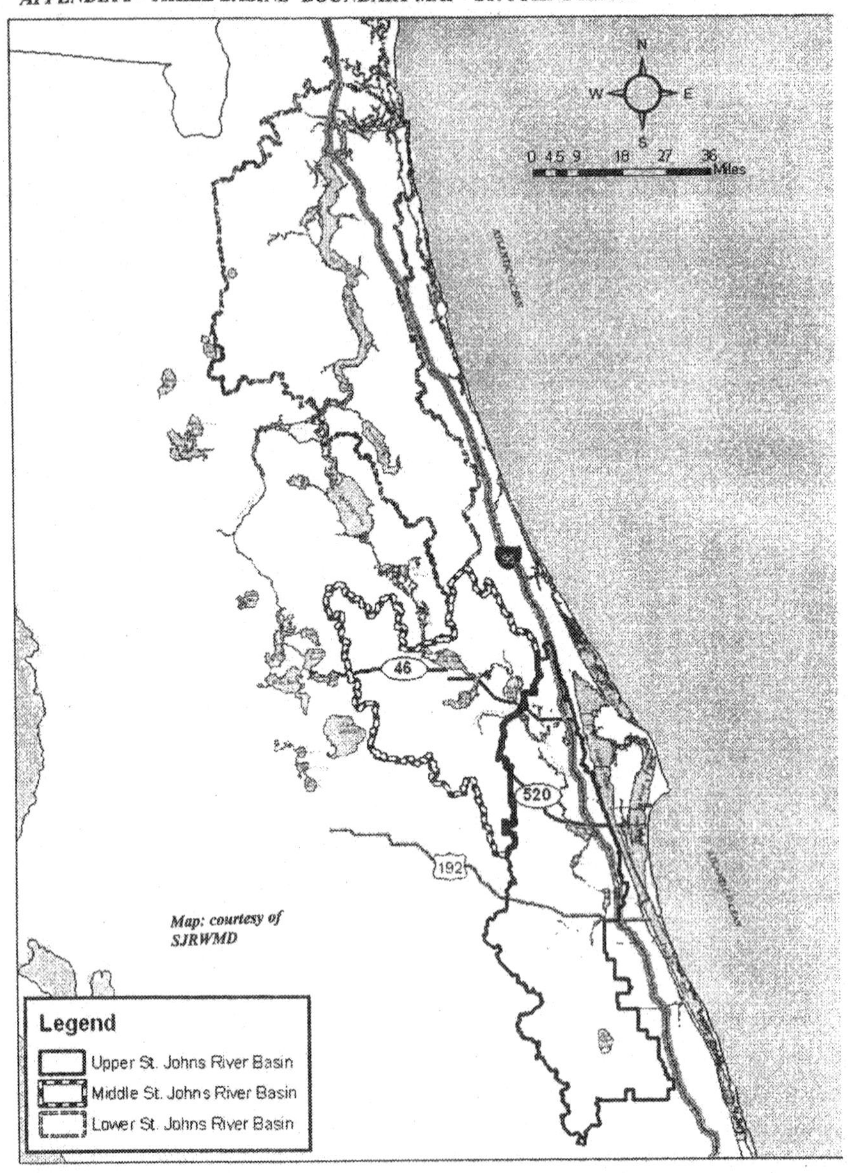

APPENDIX 3 - AMERICAN HERITAGE RIVER DESIGNATIONS

1. St. Johns River in Florida
2. Black and Woonasquatucket rivers in Massachusetts and Rhode Island
3. Connecticut River in Connecticut, Vermont, New Hampshire and Maine
4. Cuyahoga River in Ohio
5. Detroit River in Michigan
6. Hanalei River in Hawaii
7. Hudson River in New York
8. Lower Mississippi River in Louisiana and Tennessee
9. Upper Mississippi River in Iowa, Illinois, Minnesota, Missouri and Wisconsin
10. New River in North Carolina, Virginia and West Virginia
11. Potomac River in Maryland, Pennsylvania, Virginia, West Virginia and Washington, D.C.
12. Rio Grande River in Texas
13. Upper Susquehanna and Lackawanna rivers in Pennsylvania
14. Willamette River in Oregon

APPENDIX 4 - RIVER LAKES CONSERVATION AREA RECREATION MAP

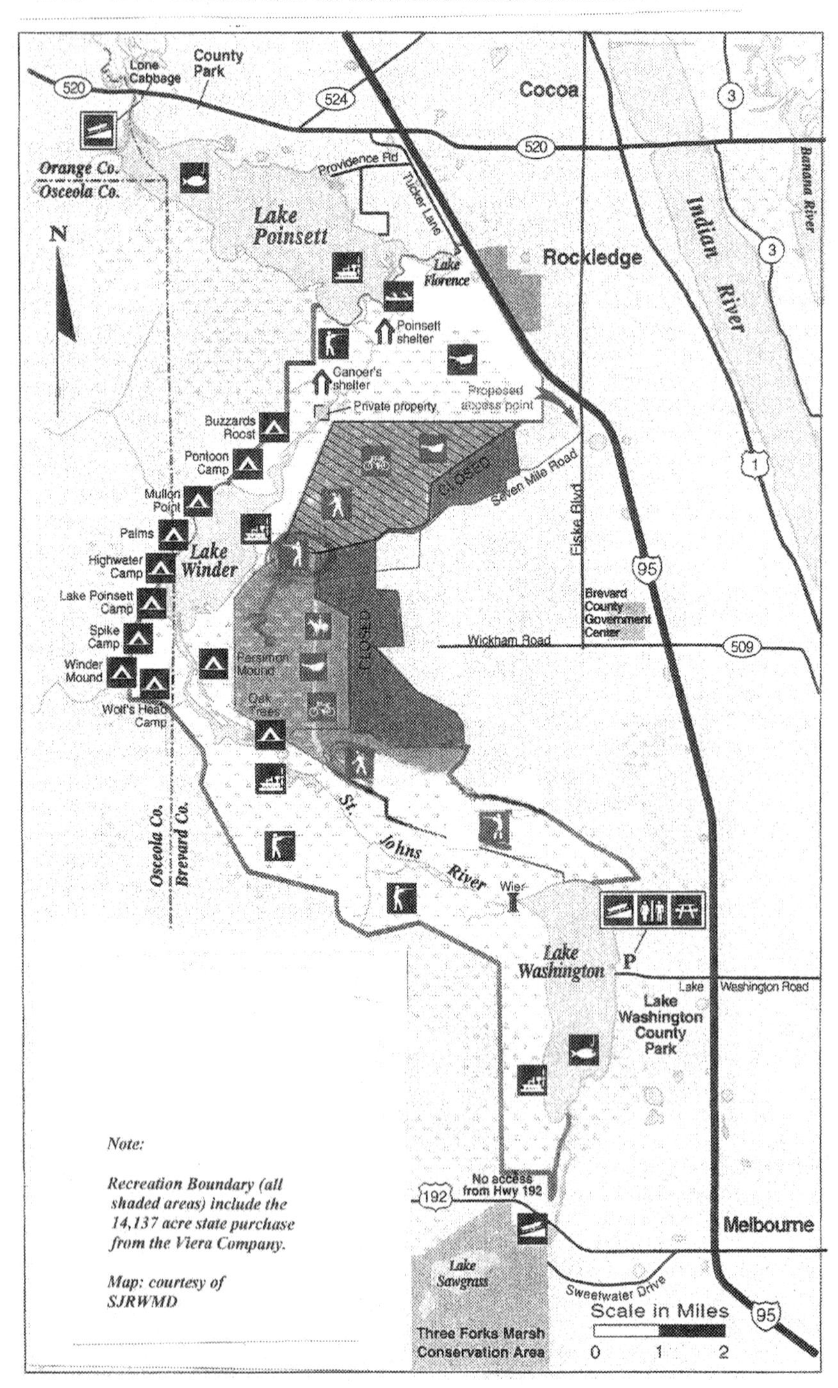

CPSIA information can be obtained at www.ICGtesting.com
Printed in the USA
LVOW04s1914191015

458908LV00012B/49/P

9 781598 582390